KYOTO AREA STUDIES ON ASIA
CENTER FOR SOUTHEAST ASIAN STUDIES, KYOTO UNIVERSITY
VOLUME 6

*The Political Ecology of
Tropical Forests in Southeast Asia*

KYOTO AREA STUDIES ON ASIA
CENTER FOR SOUTHEAST ASIAN STUDIES, KYOTO UNIVERSITY

The Nation and Economic Growth: Korea and Thailand
YOSHIHARA Kunio

One Malay Village: A Thirty-Year Community Study
TSUBOUCHI Yoshihiro

Commodifying Marxism: The Formation of Modern Thai Radical Culture, 1927–1958
Kasian TEJAPIRA

Gender and Modernity: Perspectives from Asia and the Pacific
HAYAMI Yoko, TANABE Akio, TOKITA-TANABE Yumiko

Practical Buddhism among the Thai-Lao: Religion in the Making of a Region
HAYASHI Yukio

The Political Ecology of Tropical Forests in Southeast Asia: Historical Perspectives
LYE Tuck-Po, Wil DE JONG, ABE Ken-ichi

KYOTO AREA STUDIES ON ASIA
CENTER FOR SOUTHEAST ASIAN STUDIES, KYOTO UNIVERSITY
VOLUME 6

The Political Ecology of Tropical Forests in Southeast Asia:

Historical Perspectives

Edited by

LYE Tuck-Po
Wil DE JONG
ABE Ken-ichi

Kyoto University Press

First published in 2003 jointly by:

Kyoto University Press
Kyodai Kaikan
15-9 Yoshida Kawara-cho
Sakyo-ku, Kyoto 606-8305, Japan
Telephone: +81-75-761-6182
Fax: +81-75-761-6190
Email: sales@kyoto-up.gr.jp
Web: http://www.kyoto-up.gr.jp

Trans Pacific Press
PO Box 120, Rosanna, Melbourne
Victoria 3084, Australia
Telephone: +61 3 9459 3021
Fax: +61 3 9457 5923
Email: info@transpacificpress.com
Web: http://www.transpacificpress.com

Copyright © Kyoto University Press and Trans Pacific Press 2003

Set by digital environs Melbourne: enquiries@digitalenvirons.com

Printed in Melbourne by BPA Print Group

Distributors

Australia
Bushbooks
PO Box 1958, Gosford, NSW 2250
Telephone: (02) 4323-3274
Fax: (02) 9212-2468
Email: bushbook@ozemail.com.au

UK and Europe
Asian Studies Book Services
Nijenrodeplantsoen 104
3554 TT Utrecht, The Netherlands
Telephone: +31 30 289 1240
Fax: +31 30 289 1249
Email: marie.lenstrup@planet.nl
Web: http://www.asianstudiesbooks.com

USA and Canada
International Specialized Book
Services (ISBS)
5824 N. E. Hassalo Street
Portland, Oregon 97213-3644
USA
Telephone: (800) 944-6190
Fax: (503) 280-8832
Email: orders@isbs.com
Web: http://www.isbs.com

All rights reserved. No production of any part of this book may take place without the written permission of Kyoto University Press or Trans Pacific Press.

ISSN 1445–9663 (Kyoto Area Studies on Asia)
ISBN 4–87698–453–0 (hardcover)
ISBN 1–87684–354–3 (softcover)

National Library of Australia Cataloging in Publication Data

The political ecology of tropical forests in southeast Asia.

Bibliography.
Includes index.
ISBN 4 87698 453 0
ISBN 1 876843 54 3 (pbk.).

1. Rain forests - Economic aspects – Asia, Southeastern.
2.Forests and forestry – Asia, Southeastern. 3. Rain forest ecology – Asia, Southeastern. I. Abe, Ken-ichi. II. De Jong, Wil. III. Lye, Tuck-po.

333.75095

Contents

List of Photographs vii
List of Figures viii
List of Tables ix
List of Abbreviations x
List of Contributors xi
Acknowledgments xv

1 The political ecology of tropical forests in Southeast Asia: Historical roots of modern problems 1
   Wil de Jong, LYE Tuck-Po, ABE Ken-ichi
2 Forests versus agriculture: Colonial forest services, environmental ideas and the regulation of land-use change in Southeast Asia 29
   Lesley Potter
3 Trading in the forest: Lessons from Lao history 72
   Deanna Donovan
4 The political ecology of forest products in Indonesia: A history of changing adversaries 107
   Wil de Jong, Brian Belcher, Dede Rohadi, Rita Mustikasari, Patrice Levang
5 Peat swamp forest development in Indonesia and the political ecology of tropical forests in Southeast Asia 133
   ABE Ken-ichi
6 *De facto* decentralization and community conflicts in East Kalimantan, Indonesia: Explanations from local history and implications for community forestry 152
   Steve Rhee
7 One hundred years of land use changes: Political, social, and economic influences on an Iban village in Bakong River Basin, Sarawak, East Malaysia 177
   ICHIKAWA Masahiro
8 The ecological-economics of non-sustainable development: Logging tropical forests in Southeast Asia and the Pacific 200
   Herb Thompson

9 Discourse and Southeast Asian deforestation: A case study of
  the International Tropical Timber Organization   236
  Fred Gale
10 The problem of *gaizai*: The view from Japanese forestry villages   265
  John Knight

Index   283

List of Photographs

1-1	Logged-over forests in Malaysia	10
2-1	*Baccaurea griffithii*, a common forest fruit	32
2-2	Teak *taungya* plantations in Burma	42
2-3	Assembling teak logs, Java	45
2-4	Developing a lumber concession in the Philippines	55
3-1	A *Styrax* tree tapped for benzoin resin in Laos	73
4-1	Rattan has long been used by local peoples in East Kalimantan	118
4-2	Forest products are embedded in local culture and represent local values	123
5-1	A spontaneous migrant opening a canal in peat swamps using simple tools, Riau, early 1990s	137
5-2	Large canals constructed as a government project, using heavy machinery, Central Kalimantan, late 1990s	140
8-1	Timber from a "certified" forest concession, Peninsular Malaysia	222
9-1	Pacifico-Yokohama building where the ITTO secretariat is located	242
10-1	A family timber plantation in Wakayama prefecture, Japan	267

List of Figures

	Map of Southeast Asia and the Pacific	xvii
2-1	Burma	34
2-2	Java	37
2-3	The Philippines	39
2-4	The Malayan Peninsula	48
2-5	Java and the Outer Islands	57
5-1	Landsat TM images (1989) showing land cleared by spontaneous migrants into peatlands, Riau.	139
5-2	Landsat TM images (1997) showing peatland areas in Riau, five years after the operation of the PIR project.	141
6-1	District of Malinau, East Kalimantan	154
7-1	Bakong River basin, Sarawak	179
7-2	Changing land use of Nakat village, 1947–1997	185
7-3	Change in value of black pepper and rubber	188

List of Tables

3-1	Basic indicators for selected areas of Southeast Asia, 1996	77
3-2	Significant milestones in Lao history	81
3-3	Tribute goods and exports of early major trading centers in Southeast Asia	83
3-4	Total exports from lower Laos, 1899	87
3-5	Major forest plant products entering international trade from Lao PDR, 1995/1996	89
3-6	CITES-listed wildlife products entering international trade from Lao PDR, 1998	90
3-7	Recorded export and destination of CITES-listed wildlife species from Lao DPR, 1983–1990	92
3-8	Summary of non-timber forest products exports to neighboring countries during first six months of FY 1996–1997	93
3-9	Prices in various markets for four products originating in Laos, ca. 1898	98
4-1	Categorisation of rent seeking actions used by FP producers and traders	126
5-1	Indonesian government statistics showing alleged shortage of rice-growing areas	143
7-1	Wet rice production and village expansion in Nakat	187
7-2	Number of rice fields by location, Nakat, 1995 and 1997	192
8-1	Land and forest area of small island states and dependent territories	203
8-2	Disaggregation of total economic value	206

List of Abbreviations

A.D.	Anno Domini
ca.	circa
CITES	Convention on International Trade in Endangered Species
cm	centimeter
CO_2	carbon dioxide
FP	forest products
FY	fiscal year
ITTO	International Tropical Timber Organisation
mm	millimeter
km	kilometer
ha	hectare
m	meter
ppm	parts per million
a.s.l.	above sea-level
NGO	non-governmental organisations
PIR	*Perkubunan Inti Rakyat* ("Nuclear Estate and Small-holders Project")
VOC	Dutch East India Company

List of Contributors

Abe Ken-ichi
Associate Professor of the Japan Center for Area Studies (JCAS), National Museum of Ethnology. His concern has gradually shifted from natural science to social science and Area Studies. His research areas are Sumatra and Kalimantan, focusing on eco-history of peat swamp forests, North Vietnam, Mekong River regions, and also Yunnan. Publications include *Cari Rezeki, Numpang, Siap: The Reclamation Process of Peat Swamp Forest in Riau, Sumatra* (1997), *Forest History in Yunnan (1): Tibetan God-Mountain and Its Protected Forest in Jungden* (1997, in Japanese), *Forest History in Yunnan (2): Privately Owned Forests and Eucalyptus Plantation in Han-dominated Basins* (1997, in Japanese).

Deanna Donovan
Formerly a fellow in the Environment Studies Group, East-West Center, Honolulu, and USAID Regional Forestry Advisor for Asia, she was a consultant for the Asian Development Bank, the Dutch Government, and TROPENBOS and has worked for the International Institute for Environment and Development in London. Her research topics include the effects of globalization on forest use, the adaptation of natural resource valuation techniques for community use, and the motivating factors in household exploitation of non-timber forest products. Her current geographic focus is Southeast Asia though previous work has addressed forestry and conservation issues in almost all countries of tropical and subtropical Asia. Publications include *Strapped for Cash: Asians Plunder their Forests and Endanger their Future* (Asia Pacific Issues Series 39, East-West Center, 1999); *Policy Issues in Transboundary Trade in Forest Products in northern Vietnam, Lao PDR and Yunnan, PRC* (1998); and *Development Trends in Vietnam's Northern Mountain Region*.

Fred Gale
Lecturer in the School of Government, University of Tasmania, Australia, where he teaches public policy and political economy. His research interests focus on developing an ecological political economy approach to nature, production, and power relations and to elaborating a selective

theory of international trade. He is also researching the potential of ecocertification and labeling schemes to promote ecological sustainable forestry in the Asia-Pacific He is author of *The Tropical Timber Trade Regime* (Macmillan Press, 1998) and co-editor of *Nature, Production, Power: Towards an Ecological Political Economy* (Edward Elgar, 2000).

ICHIKAWA Masahiro
Ph.D. candidate at the Graduate School of Human and Environmental Studies, Kyoto University. He has conducted research for six years in Sarawak, Malaysia, and recently conducted short-term studies in North Sulawesi and the Dominican Republic. Published papers include "Swamp Rice Cultivation in a Village of Sarawak: Planting Methods as an Adaptation Strategy," *Southeast Asian Studies* 38(1), 2000 and "Transformation of Shifting Swamp-Rice Cultivation in an Iban Village of Sarawak, Malaysia," *Southeast Asian Studies* 38(2), 2000 (both in Japanese).

Wil de Jong
Scientist at the Center for International Forestry Research, stationed in Indonesia in the Programme of Forest Products and People. Trained as a forester, he has done extensive field research on socio-cultural, technological, and economical aspects of agriculture and the use of forests among rural groups in tropical moist forest regions in Indonesia and Peru. He has also coordinated major research projects in Zimbabwe and Bolivia. His recent work has concentrated on institutional organization of the extraction and trade of forest products in Zimbabwe and Bolivia, and on how decentralization and national legislation related to forests affect forest product economies at the forest edge. Besides numerous papers in refereed journals and chapters in books, he is co-editor of *Contributions of NTFP to Socioeconomic Development* (Special Issue of the *International Tree Crop Journal*) and *Secondary Forests in Asia: Their Diversity, Importance, and Role in Future Environmental Management* (Special Issue of the *Journal of Tropical Forest Science*, 2001), and author of *Forest Products and Local Forest Management in West Kalimantan, Indonesia: Implications for Conservation and Development* (Tropenbos Kalimantan Series 6, 2002).

John Knight
Lecturer at the School of Anthropological Studies, Queen's University Belfast. He has carried out field research in mountain villages on the Kii

Peninsula in western Japan, and has written on a variety of topics to do with rural Japan, including changing human relationships to the forest. Recent publications include: "From Timber to Tourism: Recommoditizing the Japanese forest," *Development and Change* 31 (2000), "Monkeys on the Move: The Natural Symbolism of People-Macaque Conflict in Japan," *Journal of Asian Studies* 58 (1999), "A Tale of Two Forests: Reforestation as Discourse in Japan and Beyond," *Journal of the Royal Anthropological Institute (MAN) n.s.* 3 (1997). He is the editor of *Natural Enemies: People-Wildlife Conflicts in Anthropological Perspective* (2000; Routledge).

LYE Tuck-Po
Attached to the Centre for Environment, Technology and Development, Malaysia (CETDEM). An environmental anthropologist by training, her research interests include protected areas management, conflicts between conservationists and local communities, environmental ideas and histories, and forestry and biodiversity policies. She has edited *Orang Asli of Peninsular Malaysia: A Comprehensive and Annotated Bibliography* (Center for Southeast Asian Studies, Kyoto University, 2001) and has published articles on the landscape perceptions of the Batek of Pahang, Malaysia. She has written *Changing Pathways: Forest Degradation and the Batek of Pahang, Malaysia* (submitted to LexingtonBooks, Lanham, MD).

Lesley Potter
Associate Professor, Department of Geographical and Environmental Studies, Adelaide University. She has been conducting research on Southeast Asian issues, both historical and current, for the past 19 years. Her main focus is Indonesia, in many provinces of which she has undertaken field and library/archival research. She has also done work in Malaysia, Philippines, Vietnam, Laos, and Thailand. Before turning to Southeast Asian issues, she taught at the University of Guyana (South America). She is currently studying how decentralization in the "reformasi" period affects Indonesian forests and estate crops. Publications include *In Place if the Forest: Environmental and Socio-economic Transformations in Borneo and the Eastern Malay Peninsula* with Harold Brookfield and assisted by Yvonne Byron (Tokyo: United Nations University, 1995) and "The Effects of Indonesia's Decentralisation on Forests and Estate Crops in Riau Province: Case Studies of the Original Districts of Kampar and Indragiri Hulu" with Simon Badcock (Case Studies 6 & 7, Case Studies on Decentralisation and Forests in Indonesia, CIFOR, 2001).

Steve Rhee
Ph.D. candidate at the Yale University School of Forestry and Environmental Studies. His research focuses on the socio-political dynamics of natural resource management in areas of high conflict in East Kalimantan, Indonesia. His multi-sited research attempts (1) to track the flow/distribution of information from center-periphery and vice versa, (2) to map decision-making processes regarding natural resource management, and (3) to examine how various sites of political power influence each other. His theoretical interests include political ecology, environmental anthropology, and the sociology of knowledge. Since his involvement in Indonesia in 1996, he has carried out collaborative research with the Center for International Forestry Research, conducted policy analysis for the Wildlife Conservation Society, and worked as a Program Advisor for CARE.

Herb Thompson
Professor of Economics at the American University in Cairo, Egypt. His main research areas are Economic Development and Resource Economics and Computer-mediated Teaching/Learning. With particular reference to the "Political Economy of Forestry, Logging and Timber Industries in Southeast Asia and the Pacific Region", he has researched extensively in Bangladesh, India, Indonesia, Malaysia, Papua New Guinea, the Philippines, Solomon Islands, and Vanuatu. A selection of main publications include: "Crisis in Indonesia: Forests, Fires and Finances," *Electronic Green Journal* Issue 14, Spring (2001), "Social Forestry: An Analysis of Indonesian Forestry Policy," *Journal of Contemporary Asia* 29(2):187-201 (1999), "Philippine Forests: The Trees are Gone, Where's the Wood?" *Antepodium Electronic Journal*, September (1997), and "Culture and Economic Development: Modernisation to Globalisation," *Theory and Science* 2(2), Fall (2001).

Acknowledgements

The chapters in this volume began as papers given at a Japan Center for Area Studies (JCAS)/National Museum of Ethnology symposium bearing the same name (Osaka, November 28–30, 2000). Their final shape reflects the fruitful discussions shared by all at the symposium. Foremost, we would like to thank the Museum, especially its director Ishige Naomichi as well as the director of JCAS Matsubara Masatake for providing symposium space and facilities and for overall hospitality. We are most grateful to the Japan Center for Area Studies symposium secretariat, especially Noriko Iizuka, not only for organising the symposium but for shepherding this book through the production process.

Discussions at the symposium were enlivened by the presence of a number of attending scholars and we would like to acknowledge their participation here. Our primary thanks go to Sanit Aksornkoae (Kasetsart University, Thailand), Michael Leigh (Universiti Malaysia Sarawak), Kimura Hideo (University of Tokyo), and Masuda Misa (University of Tsukuba), whose symposium papers provided important material for comparative reflections.

For their thoughtful commentaries on the various papers, we would like to thank Tanaka Koji (Center for Southeast Asian Studies, Kyoto University), Sato Jin (University of Tokyo), and Uchibori Motomitsu (Toyo University of Foreign Studies). For moderating the various sessions and contributing commentaries, we are indebted to Oikawa Yosei (University of Agriculture and Technology), Soda Ryoji (Hiroshima University), Ishikawa Noboru (Center for Southeast Asian studies, Kyoto University), Fujita Wataru (Kyoto University), Yamada Isamu (Center for Southeast Asian Studies, Kyoto University), and Akimichi Tomoya (Research Institute for Humanity and Nature, then of the National Museum of Ethnology). Other useful commentaries were made by Andi Amri (Kyoto University), Furukawa Hisao (Kyoto University), Hyakumura Kimihiko (Institute of Global Environmental Strategies), Kanazawa Kentaro (University of Tokyo), Ogino Kazuhiko (University of Shiga Prefecture), and Okuda Toshinori (National Institute for Environmental Studies).

During the preparation of this volume, Wil de Jong and Lye Tuck-Po revisited the Museum in 2001 and 2002 as JCAS Visiting Fellows and we would like to thank Oshikawa Fumio, director of JCAS for extending the invitation. In working up these papers for publication, we have received useful feedback from Carol Colfer and David Kaimowitz at CIFOR, and we would like to thank them, as well as two anonymous reviewers. As editors, we are also grateful to the various authors of these chapters, who responded to calls for revision—and usually with good speed!

Finally, we would like to thank Tetsuya Suzuki of Kyoto University Press for his kind assistance with the publication of this volume and the Japan Society for Promotion of Science for a grant-in-aid for publication of scientific research results, which made this publication possible.

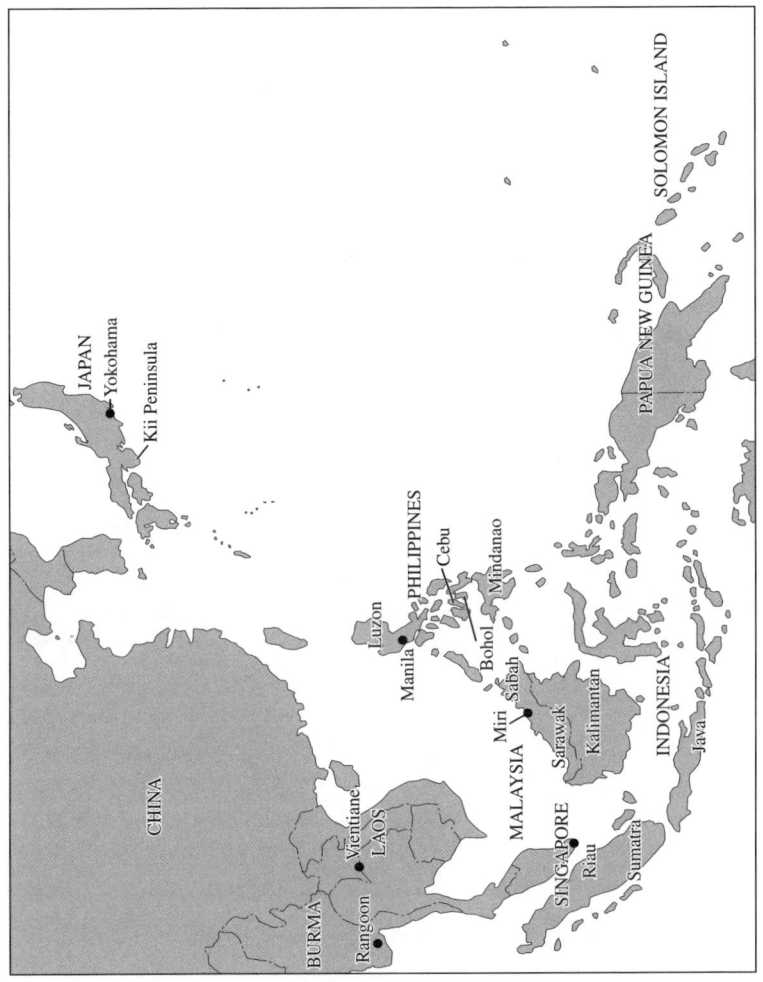

Southeast Asia and the Pacific

1
The Political Ecology of Tropical Forests in Southeast Asia: Historical Roots of Modern Problems

Wil de Jong, Lye Tuck-Po, and ABE *Ken-ichi*

Introduction

Forest change and conversion in Southeast Asia, and changes in the reproduction and supply of forest resources, can be attributed to a fairly narrow number of causes. Swidden agriculturists have long used forestlands for agriculture and settlement. However, since at least the second half of the 20th century, logging and related industrial timber plantations have been the main causes of forest cover change in the region's countries. Recent estate crop production, dominated by oil palm, is leading to clearance and conversion of sizable tracts of forest. The most recent cause of irreversible change on a large and dramatic scale is fire, which in turn is closely related to logging, land clearance, and modern estate crop production (Dennis *et al.* 2001; Glover and Jessup 1999). All these are contemporary processes with complex and heterogenous long-term implications but they have in common an underlying economic interest in the forests of Southeast Asia. This is an interest claimed by different groups and sectors in societies and these claims are marked by uneven access to knowledge, technologies, status, influence, and power.

There are deeply embedded historical roots to these problems. The history of swidden agriculture's expansion, for instance, includes the pushing out of weak groups from coastal regions or fertile valleys by stronger groups (for example, Schmidt-Vogt 2001). Plantation development goes all the way back to the 19th century, although its relative contribution to forest conversion has shifted from one country to the next, and from region to region within countries. The logging industry, although reaching a boom during the 1980s and 1990s, did have its late 19th century forerunners in the same regions.

Tropical forest issues have risen to high political prominence since the 1980s, when the logging boom in Malaysia and Indonesia was in full swing. This attention was part of the worldwide concern for the destruction of the earth's environments, which was provoked by growing fear of threats to resource supplies and biodiversity and to some degree by advances in the conservation agendas of the Northern countries.[1] In the wake of this perceived onslaught on tropical forests, researchers and activists alike started to question the politics behind the events that were taking place, and that had led to this widespread degradation of Southeast Asia's forests (for example, Gillis and Repetto 1988).

Researchers inspired by political economy approaches, who had contributed studies on the social inequities caused by deforestation, contributed to a new mode of analysis that is now known as political ecology. As a scholarly undertaking, political ecology identifies a type of analysis of the links between natural resource use and the political economy of that use. This is not a new way of examining issues, as shown, for instance, by studies of the rubber boom in South America around 1900 (Hardenburg 1912). However, in view of the late 20th century fate of the rural poor living under expanded commercialized resource regimes in remote corners of the world, global conservation concerns, and the modern sustainable development discourse, political ecology has become a more critical tool of analysis and its prominence has grown.

The papers in this volume contribute detailed cases and analyses concerning some, if not many, of the most recurring issues in the political ecology of Asia's forests. An added advantage is that this volume gives more explicit attention to the ecology of the region than some other recent studies (see, for example, Hirsch and Warren 1998; Li 1999; Parnwell and Bryant 1996). A number of environments and resource regimes are covered here: from upland to insular Southeast Asia, from dipterocarp to peat swamp forests, and from "major" to "minor" forest resources and products. It will be clear that, amidst the diversity, there are some common themes and issues in the exploitation and benefit-capture of forests. The orientation is towards issues. Central to our concern is the role of history in environmental relations. The approach taken in this volume is to begin with conditions in Southeast Asia today and inductively trace causes and processes to past events, precedents, or precursors. As a group, the contributors to this volume share a background of integrating practical work with conceptual discussions; this practical experience informs the examination and presentation of material. Another notable feature is the multi-disciplinary background of the contributors; though trained in conventional fields of inquiry, like ecology, economics, forestry, anthropology, and political science, their experiences of "real world" issues often lead them to go beyond disciplinary boundaries in their approaches and analyses.

This introductory chapter selectively reviews some key political ecology topics that set the stage for the chapters that follow. We do not review the development of political ecology, or its conceptual promises and limitations, as that has been done elsewhere (Bryant 1992; Bryant and Bailey 1997; Greenberg and Park 1994; Peet and Watts 1996). Rather, the concern is with those lineaments of political ecology that this volume most fittingly contributes to. In the next section, we critically review the definitions and components of political ecology that frame discussions in the book. Then, following a review of current issues and topics in Southeast Asian political ecology, we examine some of the relevant historical issues. Section three will highlight the contribution that each of the chapters in this volume makes to the broader issues at stake. Section four will take these conclusions back to the issue of political ecological analysis in general.

The Scope of Political Ecology

The field of political ecology has been referred to as a method of analysis, rather than a unified scientific discipline or sub-discipline, which is usually characterized by a set of related ideas, premises, and theories. However, simply dismissing political ecology as a method of analysis (as in, for example, Peluso 1992a) does not do justice to the commonalities of themes that are of interest to most discussions grouped under this topical category. Probably political ecology can be described as a not clearly defined "confluence" (Peet and Watts 1996: 6) of related sub-disciplines. For some, political ecology aims to integrate both political and economic inquiry with orthodox ecology, in the sense that a clear connection is being sought between what happens at the political level related to the economics of natural resources and how that affects the use and therefore ecological functioning of those resources (Blaikie and Brookfield 1987; Greenberg and Park 1994). Such a unified analysis requires integrating social and biological sciences in a more sophisticated way than is commonly possible by analysts working from any single discipline. Although this characterization truly reflects both the political and the ecological, most political ecology discussions are political economy studies of natural resource use, where the *ecology* of those resources (as understood by biologists) is secondary to the discussion. As such, conventional political ecology better fits the definition offered by Peluso (1992a: 51): "political-economic analyses of the environment that incorporate some discussion of the actions of the resource users and their linkages to the broader processes that structure the social and physical environments in which they act."

As this is a relatively new academic trans-discipline with diffuse beginnings, a number of political ecology practitioners have tried to come up with definitions (for example, Peluso 1992a; Mayer 1996), leading to quite a diverse selection of meanings. In this volume, we understand political ecology as: *a collective name for all intellectual efforts to critically analyze the problems of natural resource appropriation and political economic origins of resource degradation, be they for the purpose of academic study or practical applications.* In other words, political ecology is concerned with the political dimensions of natural resource use, and the subtleties of those politics. Central to these efforts is the "political" as primary in the analysis of resource use. This definition implies that in some instances the political shapes the ecological, but it is equally relevant to state that in turn the political is being shaped by nature's feedback. The definition also implies that conflicts and contestation are essential components of environmental relations. Political ecology not only looks for conflicts in resource appropriation, but assumes conflict to arise when the values of resources change. This means that conflict cannot be viewed as anomalous but as an increasingly integral dimension of human-nature interactions.

A second feature of political ecology, probably less clearly emphasized in many definitions, is the importance of identifying power relations. Issues related to power are, of course, integral to the political economy approach. Political economy, with its roots in Marxian theory (Greenberg and Park 1994), is interdisciplinary to begin with, as it tries to relate economic trajectories to mandates, political influence, and access to power of actors and agencies at different administrative levels. Implicitly such analyses are grounded in an underlying value framework, that we might call "moral economy", in which there is a concern to point out how the distribution of power shapes the unequal distribution of rights and responsibilities. Problems of distribution are entailed by, and entail, structures of legitimacy, where the deprivation of environmental goods and benefits from those who are less powerful or less well-connected politically is accepted and interpreted as "natural". As one example, a political ecology approach does not assume that poverty, or resource degradation by "poor" forest-dwellers, is an adequate explanation for problems (Peet and Watts 1996: 7–8). Rather, it aims to show how those problems are embedded in structures of inequality, and can be traced to the use and misuse of capital.

This focus on the political and on the power structures underlying resource use separates political economy and political ecology from other disciplines such as "environmental economics". Environmental economists are primarily concerned with the problems facing industrial nations, manifested (for example) in attempts to resolve forest degradation in Southeast Asia through the exchange of carbon emission rights. For political ecology, as the foregoing implies, emphasis is not on "quick fix" or narrowly technological solutions to problems (Thompson, Chapter 8) but rather on the complex relationships and characteristics of states and nations, and how social groups are differently positioned within wider political organizations. Thus, from early beginnings in recognizing the role of power, politics, and influence in environmental and resource appropriation, political ecology has evolved into a critical disciplinary conjunction that questions the foundation of society-environment relationships.

The kinds of problems that political ecology addresses are both of the "state of the world" variety (i.e., empirically grounded studies) and of problems of appropriation. These are often cases of coerced expropriation of natural resources previously controlled or managed by forest dwellers, that lead to negative impacts on their well-being or aspirations to improve patterns of resource use. In the economic sciences, the actions that are scrutinized in political ecology discussions fall under categories of rent seeking. In economic theory rent seeking has a broader meaning, as it includes obtaining profits by political power, connections, or ownership of natural resources. However, as de Jong *et al.* (Chapter 4) point out, there are different degrees of rent seeking, with "coercion" being the highest level of abuse of power, ranging in forms from cronyism and cooptation to the threat or actual use of physical violence. Political ecology inquiries trace such processes back to root sources of problems, recognizing that those linkages usually extend quite far geographically and quite far in time.

Central in such efforts is the acknowledgment that different agents and social groups, who have different resource interests or access to resources, are mutually connected through a range of social, political, and economic relationships. Such relationships may be class, ethnic, and/or gender-based. Theory and practice are, therefore, confronted by many levels of agent involvement: from local communities (for example, forest-dependent peoples including shifting cultivators, transmigrants, and other small-scale farmers) to multilateral and international institutions like the World Bank and Global Witness; from local ecological impacts to regional externalities (for example, air pollution and health degradation in Singapore, Malaysia, and the Philippines arising from fires in Kalimantan); from small-scale middlemen to national and multinational business interests and environmental groups like Greenpeace. The overarching proposition appears to be: if capital accumulation is to continue unabated, nature is to be exploited. Under the prevalent property and entitlement arrangements, this often will benefit some and harm others. If nature is to be preserved, capital accumulation requires regulation. The state is the site of struggle wherein the priorities, antagonisms, and conflicts between capital and nature are worked out. How is this done, for whom it is done, and when it will be done, is never deterministically decided.

Contemporary conditions have increased awareness of subtle linkages between different political and economic interests. Part of the task for political ecology is to sort out what are the different levels of analyses, both in space (using insights from anthropology, political science, and geography) and time (history), and to be guided by real world conditions, which other sciences can provide the tools to help examine (like soil science, ecology of species and their habitats, forestry, etc.). This is a different position from much of the extreme postmodern approach that essentially distrusts scientific knowledge (on grounds that science is a part of the authority structures that contribute to resource degradation). A political ecology position, in this respect, would hold that "conventional" science should be critiqued, but not discarded, as long as the applications of science are understood to be subject to politics, as shown (for example) by Gale (Chapter 9) in this volume.

Ultimately, political ecology asks questions about the nature of society and the nature of nature, while not forgetting that nature is not an adjunct to be exploited by society, but encompasses the reality of which society is an active ingredient. Political ecology takes a hard look at the environmental and social impacts of particular agents and their actions. The logical conclusion to a political ecology mode of explanation is that politics is also the absolutely crucial part of the mechanism of halting resource degradation.

Political Ecology And Southeast Asian Forests

One of the earlier topics that received broad publicity, even before tropical deforestation entered center stage, were the Amazon development plans of

Brazil in the 1970s and their negative effects on forests and indigenous peoples (Browder and Geoffrey 1997). Apart from earlier accounts that discussed the atrocities of the rubber boom (Hardenburg 1912), this was among the first political ecology discussions related to tropical forests. This set of studies was later followed by critical discussions of national policies affecting colonization of the Amazonian region (for example, Bunker 1985). These policies included tax incentives, meant originally to encourage investment for land development but which often led to forest conversion for the sake of land speculation (Hecht and Cockburn 1989). Since the late 1980s, a coalition wedding together local rubber tappers with national and international activist organizations has taken up arms against the cattle ranchers who benefited from such tax exemption opportunities. As with the parallel concern with South America's native peoples or indigenes more generally (for example, Assies *et al.* 1999; Conklin and Graham 1995), such movements aspire to save the forests for the sake of the people most directly affected by forest clearance and land dispossession. However, they were also responding, in part, to the Northern concern for the long-term survival of the tropical forests and their biological diversity.

A political ecology attention to the forests in Asia also accelerated in the 1980s. During that time, most of the logging of tropical timber took place in Indonesia and the Malaysian state of Sarawak. Conditions in these two countries came under scrutiny by the international community, especially in view of their dual status as countries with high indices of biological diversity that were also among the largest producers of tropical timber in the world. (The Philippines had already gone through most of its forest conversion by that time, but that fact was largely unnoticed when it was happening; by the 1980s, little was left of the original forests there.) An apt symbol of the struggle to save the forests from destruction were the anti-logging blockages erected by Penan hunter gatherers of Sarawak in the mid-1980s (Gale, Chapter 9); however, local opposition to logging was widespread and had in fact been mounting for years (Anon. 1992; Hong 1987). Conflicts between logging and plantation companies and the indigenous inhabitants remain central in critiques of forestry programs in Sarawak and Kalimantan, the Indonesian part of Borneo (Peluso 1992a; Mayer 1996; Rhee, Chapter 6; de Jong *et al.* Chapter 4).

Related to the above have been discussions on the political ecology of non-timber (or minor) forest products, brought to international attention by researchers like Peluso (1983, 1992a, 1992b, 1992c) and Dove (1993, 1996). The late 1980s saw conservationists advocate the commercialization of non-timber forest products to advance rural development and conservation agendas (see, for example, Oldfield and Alcorn 1991). It seems clear that such projects were strongly influenced by modernization theory: i.e. they were backed by the view that if the poor have an economic stake in the rainforest, they will be more developed and less inclined to destroy the forest. Even if that premise were true, it is arguable whether the people could retain control of their resources, let alone benefit from them. Dove (1993), taking a position subsequently adopted by others (Neumann and Hirsch 2000; de Jong *et al.* Chapter 4), argued that poverty was not the root problem and that successful commercialization would inevitably lead to the control of the trade by well-connected business elites, thus leaving little of the profits to primary collectors. A related set of critical discussions saw problems with commercialization and benefit capturing of forest products (for example, Peluso 1992b, 1992c).

Reviews of the logging *industry* as a destroyer of tropical forests began about the late 1980s when criticisms were advanced about the highly preferable conditions under which forest companies were given access to forest resources, while nation-states lost important income from the opportunities these resources gave (Repetto and Gillis 1988). Much subsequent research has focused on investigating the complicated entanglements of logging and plantation companies with national, regional, and local elites (see Cooke 1999). These sorts of relationships put entrepreneurs in a favorable economic position. Frequently, granting logging concessions is part of the process by which politicians reward "cronies", as seen, for example, in Indonesia, where the military has been strongly implicated in logging operations and recognized as a critical supporter of the Suharto regime (Ross 1996). Entrepreneurs may have the ability to pay their way out of restrictive regulations (for example, those promoting 20- or 30-year cycles of rotation, or banning logging in hilly areas beyond a certain slope level) or to escape penalties for illegal logging practices. In principle, as is widely stated, nation-states in the region often have a good framework of environmental rules and regulations. Failure to implement these regulations has less to do with inadequate enforcement than with the political entanglements and structures that hinder the management and control of logging (for example, Ross 1996; Barr 1998).

A contentious issue continues to be the "national development" versus the "enrichment of the selected few" dichotomy. One persistent issue discussed in many studies is the cost of tropical deforestation or forest degradation to world society, in terms of the loss or extinction of biological diversity (Thompson, Chapter 9; Wilson 1988). Around the early 1990s, however, and reaching a peak in the 1992 "Rio Summit" (the United Nations Conference on Environment and Development), some governments argued that the preservation of species was overridden by gains in national development in their countries. In its starkest form, this is a debate between saving the forests for conservation purposes and eradicating poverty in developing countries.

These related issues—the plight of forest dwellers, projects to alleviate poverty and promote community entrepreneurship of forest products, and the machinations of the logging industry—became central to a whole body of political ecology discussions. Political ecologists, along with conservationists and rights advocates, argued that logging, land clearance, and estate crop plantations were not a sustainable or even legitimate method of converting natural resources into capital for social and economic development. Society at large could not benefit at the expense of marginal groups. Further, the sustainability of the resources was being jeopardized. One problem is that the system in its current state puts the logging industry, its players and political patrons, out of the rule of law, and enables a relatively small group of players to ignore customary rights of communities, sustainability issues, and even national concerns to retain adequate amounts of forest cover. In an early line of counter-attack, activists and researchers argued that the loss of forest contributes to rural poverty, since this deprived communities of resources for daily sustenance and forced them into a more systematic pattern of market dependance—which they could not control (see Zahid 1990). Such issues of inequity were first brought to public attention by NGOs like Survival International in London and subsequently examined in studies like Hurst (1990).

While encroachment by business interests into tropical forests was taking place, government agencies were not unaffected by such criticism. One governmental response that developed from about the late 1970s onwards was to resurrect the old hare (see below) of blaming swidden agriculturists for the progressive destruction of the forests. Studies were commissioned to analyze the persistent "problem" of swidden agriculture and propose solutions (for example, Kartasubrata 1993; Weinstock and Sunito 1989; Hatch 1982; Hatch and Lim 1978). Quite a few programs were developed to address this "problem" (critiqued by, for example, Weinstock and Sunito 1989; de Jong 1997; Potter and Lee 1998). At the same time, more and more researchers systematically questioned this official discourse against swidden agriculture and exposed its ideological roots (Dove 1983;

Sunderlin and Resosudarmo 1996). Not the least, studies that looked at integrated resource management and forest management practices among swidden agriculturists (Poffenberger 1990; Padoch and Peters 1993) have ultimately led to widespread proposals for devolved forest management by local communities.

In the aftermath of the timber boom, a new wave of resource appropriation came along, this time not for the forest resource itself, but rather for the land to be used for "mega-projects" like hydro-electric dams and estate crop or industrial timber plantations. As with similar processes a century earlier (see below), these programs may have more serious consequences for the future of forests and forest-dependent communities. Under a logging regime, it is at least theoretically possible to enforce

Photo 1-1. When logged-over forests have a chance to regenerate, they may continue to provide substantial environmental services to local peoples. Here Batek hunter-gatherers of Pahang, Peninsular Malaysia, set off to check on the condition of fruiting trees. *Photo by Lye Tuck-Po*

spatial and temporal limits to cutting cycles, in order to ensure adequate amounts of time for logged-over forest patches to regenerate. Forest communities may find their resources depleted and landscapes culturally impoverished, but they themselves are not necessarily displaced from territories (see Lye 2000). Under plantation and dam-building regimes, however, resettlement of peoples and irreversible transformation of forests are necessary preconditions. At best, as detailed by Abe in Chapter 5, communities may become laborers working for plantation companies; they lose not only their lands but also their rights to those lands, cultural and spiritual assets, and the basis for identity, cultural continuity, and autonomy.

There are many examples from Indonesia where larger companies, often supported by local government officials, police, and the military coerced local villagers to accept plantation projects in their backyards, who then often were left with the choice to either participate in the projects or move elsewhere (Dove 1985; Mayer 1996; Potter and Lee 1998). Estate crop development is long-standing in the region but it was only during the 1997–1998 forests fires in Indonesia that caused widespread health and pollution hazards did it become apparent that plantation development was a major contributor to those fires. They most likely had been a major contributor to similar fires that occurred earlier (for example, 1980s and early '90s) but at that time they were not singled out as such.

Historical Roots of Contemporary Issues

The History of Forest Products Trade and Organization of States
Some treatises on political ecology give overriding primacy to the structures of power introduced by colonial regimes and that persist in the modern development discourse (for example, Peet and Watts 1996). However, as Donovan (Chapter 3) and de Jong *et al.* (Chapter 4) demonstrate, key aspects of the political economy of Asia's tropical forests date from before the arrival of European powers in this part of the world. One element shaping the organizational structures of small states (sultanates and kingdoms) was the struggle to exert control over the forest products trade—the major source of revenue and wealth until the colonial era and over which wars were fought. Peluso (1983) and Sellato (2001), for instance, show how several sultanates in Borneo emerged because of their manipulation of the China trade, which is estimated to be as old as the third century A.D. (Peluso 1983). The same was true for other places

like Peninsular Malaysia where early state formations were enabled by control over resources from land and water, as well as over key ports and shipping lanes (L. Andaya 1975; Andaya and Andaya 1982).

The economic-political organization that resulted from struggles over territories persisted until very recent times. Initially, it was the social organization of collection that led to the founding of small states, and the need for efficient collection mechanisms affected how those states were organized. In addition, the spatial organization of inland groups became largely influenced by the desire and efforts deployed to appropriate forest products, minerals, and other lucrative resources like birds nest caves. In East Kalimantan, for instance, the territorial organization of the interior groups today can be explained to an important degree by processes resulting from conquered access to resources (de Jong *et al.* Chapter 4). Many of the contemporary hill tribes were pushed out of fertile regions by stronger groups and had to retreat to remoter mountainous regions (Schmidt-Vogt 2001). What has appeared to be intrinsically "natural" upland adaptations of many groups now on closer inspection can be attributed to forced retreats.

Ethnic formation and territorial locations, then, can be attributed to developments in the exploitation of forest resources. How these processes happened varied from one place to another, and from one political regime to another. In Borneo, Dayak groups were displaced from river valleys by Malay groups, which may have been immigrants or "converted" Dayaks (King 1993). Malay encroachment was a result of the search for agricultural lands but also for the forest products that formed the basis of sultanates' economic wealth (King 1993; Sellato 2001). These various local histories, with different implications for understanding relationships among groups and processes contributing to ethnic formation, demonstrate the link in Southeast Asia between the search for wealth through forest resources that had some wider demand, and the economic and political organization of the region (Li 1999).

The Historical Roots of Problems with Forest Policies and Legislation

Although colonial regimes did not introduce a political economy of trade to the region, there is no doubt that they did bring with them a "modernist" perception, knowledge, and management of forests and forest resources.

Two of the main economic activities engaged by colonial powers were the appropriation of the forest products trade for the lucrative European market and occupation of territories for estate crop production. Changes effected by the latter (Cant 1973) practically reshaped the forest landscape.

The scale of these plantations meant that it was important to find "new" supplies of good land; thus, colonial powers were brought "into the forest" in a way that older, precolonial regimes had been unable or had no justification to do. A common colonial view throughout the region was that forests had no value unless converted first to agricultural land (Lim 1973). However, as Potter (Chapter 2) discusses, the value accorded to agriculture over forestry was subject to changes in commercial, environmental, and political concerns, including the status and influence of forestry departments in colonial bureaucracies.

The establishment of forestry departments in the late 19th century served to regulate the development of economic activities. These agencies, their policies and actions, have had long-lasting influences. Potter argues that, although the primary concern of forestry agencies was economic—to aid in the expansion of colonial regimes—they did to some extent have conservation concerns. These concerns often could not compete with the larger economic interests of colonial powers (Edmund and Wollenberg 2001). Double standards were rife, as Potter notes. Notorious was the widespread condemnation of swidden agriculture for its perceived, but largely unproven, negative environmental impact. Since the late 19th century environmental theories that predicted negative impacts of deforestation on local rainfall patterns and downstream water regulation served to condemn swidden agriculturists but, ironically enough, not the colonial estate-based planters. Although such theories had their influence on colonial foresters' attitudes towards swidden agriculturists, another reason could be that foresters were also competing with the swiddeners for prime forestlands. Foresters had a duty both to manage resource stocks of timber *and*, to justify their work in a commercially-minded colonial economy, to make timber economically productive. Forest communities, whether in Malaysia, Indonesia, or the Philippines, were easily targeted for official criticism because they were living in the lands that contained the supplies of timber coveted by foresters and entrepreneurs. This denunciation of swidden agriculture has become entrenched in state-centric ideologies, being revived periodically during times of perceived crises in the forestry sector. In the 1950s, for example, the classic study by Conklin (1957) was a scholar's response to official denunciation of swidden agriculture in the Philippines. As we saw earlier, this line of attack was revived in the 1970s and 1980s and, more recently, during the 1997–1998 forest fires.

As an instrument of control, environmental legislation tends to focus on some aspects of a problem to the neglect of others. How those problems are selected is often due to politics and economics rather than science. The case of swidden agriculture is a good instance of this (see Dove 1983). As several authors (for example, de Jong *et al.* 2001) have pointed out, still persisting are those legal instruments developed during the early decades of colonial forestry, which condemn or outrightly prohibit swidden agriculture, and policy measures that were meant to solve this "problem". These contemporary policies have for a long time succeeded in putting a large part of the blame of deforestation on swidden agriculture. While this happened they achieved to detract attention from what are now recognized to be major damaging factors to Asia's tropical forests: widespread logging and plantation development.

Social Roots of the Onslaught on Asia's Forests

Although quite a few of the problematic forest policies and legislation can be traced back to previous times, some contemporary policies and practices demand alternate explanations. As discussed earlier, international market interests in the dipterocarp species of tropical forests have since the second half of the 20th century stimulated waves of commercial investments in the forests of the Philippines, Malaysia, and Indonesia. These investments, it is now widely recognized, have been characterized by widespread collusion that for an important part explains the total lack of implementation of regulations that were meant to establish some minimal degree of sustained supply of the timber resources (see Barr 1998; Cooke 1999; Potter and Lee 1998). An important explanation of the dynamics leading to the onslaught against the forests are the interpersonal relations between the principal players involved.

A few examples demonstrate this. In Indonesia, relationships between ex-President Suharto (who held supreme power for 32 years) and his close cronies—the development of political patronage—explain how the logging boom evolved. Cronyism may have reached its height during this period, but it did not spring forth on empty soil. Some of the key interpersonal relations were developed from those that had predated Suharto's attainment of power (Barr 1998). In the deployment of power, cultural notions and images—for example, those from Javanese traditions of leadership (Abe, Chapter 5)—helped to naturalize and legitimise the process of resource appropriation.

Another example of the significance of social relationships to forest exploitation is the Sarawak case. Leigh (2000) shows the dominance of one Chinese ethnic group, the Foochow, as contractors responsible for the actual execution of the logging operations in the state. The Foochow's success in occupying this niche in the timber industry is due to some shared features that stimulated collaboration with non-Chinese players, especially the Melanau that dominate state politics. These characteristics can be explained by their collective history, beginning from when they left China at the beginning of the 20th century and their strong social bonds and particular socio-moral sub-culture while establishing themselves in Sarawak.

Such relational issues are not limited to the highest levels of power, whether today or historically. Social relationships between different ethnic groups in the remote regions of Indonesia explain some patterns of resource use. The patron client relations between dominant Dayak groups in East Kalimantan and the true hunting gathering Punan groups characterize trading dynamics and patterns of resource exploitation by the latter. Contemporary processes erode these patronage systems. In some regions the Punan now bypass their former patrons (Momberg *et al.* 1997) but in river systems like the Bahau and the Malinau the relationships may persist (Sellato 2001). The possibility of "secularizing" resource management—freeing resource appropriation from restrictive social entanglements—may seem attractive from a purely administrative point of view. However, as Rhee shows in Chapter 6, changes in relationships and dynamics do bring their own sets of problems.

Cultural Histories and Moral Economy

An overwhelming concern in the political ecology of forest destruction have been the impacts on the economic systems of the less powerful actors (for example, de Jong *et al.*, Chapter 4). This bias towards economics has been subjected to criticism; the dominance of political-economic analysis can be held as another example of the dominance of Western rationality and values. Other, non-economic, values and moral economy need to be part of political ecology discussions. Rhee's Chapter 6, for instance, demonstrates the decline of social relations, or social capital, and an increase in inter-ethnic conflict between native Dayak groups resulting from the current decentralization in Indonesia and the increased decision-making powers of provincial and district governments. He describes a history of conflict between groups that goes back to precolonial days. These conflicts, as also examined by de Jong *et al.* (Chapter 4), were often caused by outsiders' interest in local forest products and their subsequent actions to purposely use some groups to oppose others. In other words, conflicts among collector groups were part of the mechanism by which outsiders appropriated local resources.

Although these conflicts have been a recurrent historical event, this fact should not serve as a mere explanation of the conflicts described. Rather, it becomes evident from the Rhee chapter that regional development decisions, even if they have the improvement of the welfare of rural people in mind, do insufficiently consider non-economic costs and benefits (Thompson, Chapter 8; for parallel criticisms, see Armstrong 1998; Gudeman 1986). This becomes very evident in the analysis by Knight (Chapter 10) of the impact of the timber trade on Japan. This example profoundly demonstrates the predominance of power and subsequent control of decisions based on control of economic assets, as opposed to spiritual assets. This, it can be argued, is a problem that has been recognized widely by those who have come to understand the profound meaning of the differences between cultural and moral values and economic ones.

Contributions to Political Ecology

The composition of chapters presented in this volume, and their geographical scope, reflect a thematic congruence. Chapters 2 to 4 are most explicitly historical in scope: colonial forest bureaucracies (Potter), the centuries-old trade in Laotian forest products (Donovan), and the history of contestation over procurement of benefit-capture of forest products in Indonesia (de Jong *et al.*). Moving slightly away from the "grand sweep" of history, Chapters 5 to 7 then offer detailed studies of present-day conditions and the events leading up to them: the effects of decentralization on local politics in East Kalimantan, Indonesia (Rhee), the response of Iban villagers in Sarawak to a constant influx of economic opportunities (Ichikawa), and the exploitation of peat swamp forests in Riau Province and Central Kalimantan (Abe). Chapters 8 and 9 then explicitly address the problems inherent in the movement of resources and capital: a critique of the neo-classical approach to economic valuation of forests and their resources in Asia and the Pacific (Thompson) and an account of the International Tropical Timber Organization as a forum for struggles over the sustainable forest management discourse (Gale). Finally, to bring the narrative to a close, we venture slightly out of Southeast Asia, to look at the situation from the point of view of an upland Japanese timber-producing community (Knight); this provides a valuable comparative angle. The rest of this section elicits some shared contributions of this diverse group of studies.

The political ecology approach recognizes that resource use is politically inscribed and historically constituted. In tracing the movement of resources, such as the relationship between producers and consumers (see, for early demonstrations of this, Mintz 1986; Wolf 1982), it is clear that restricting analysis to national borders will not reveal the full implications of forest destruction. Historically and today, the activities of Japan in Southeast Asia have been a major factor in the changing uses of forests (Dauvergne 1997). Discussions about the political ecology of Southeast Asia are partial without considering the relationship between these communities and countries and Japan, and more broadly, between producer communities in one part of the world and consumers in another. Tracing the linkages, by examining intricate webs of discourse and rhetoric, market and forest, local and global, such as is done in this volume, is a necessary analytical approach.

Several chapters in this volume, even when limiting themselves to a single nation or regional focus, either hint at or explicitly address transboundary issues. The peat swamp development projects in Sumatra examined by Abe rely on foreign aid and capital that have been ploughed into the development of the SIJORI (Singapore-Johor-Riau) triangle, which is co-funded by Singapore and Malaysia. The domestic implications (political, social, and ecological) examined in the chapter, then, ultimately can be traced to extra-national origins. The trans-boundary issue comes back in a different fashion in the chapter by Donovan. The whole of Indo-China is so interwoven, internally and with China, that the French colonial powers did not disaggregate statistics, resulting in the lack of data on trade patterns in one specific country. This leads to a situation where research constraints reflect real world conditions that can be described as the lack of national autonomy in resource use and decision-making. The problems are now grave: as Laos emerges from isolation it finds itself in charge of one of the great remaining reservoirs of biodiversity in the region but it may not have the capacity to control resource theft and exploitation within its own borders.

The historical and ideological dimensions of this are addressed in the chapters by Potter and Gale. As Potter details, there were some broadly identifiable characteristics of the various colonial forestry approaches, namely the application of theories and practices from the French-German schools and a certain myopia in focusing on a narrow list of commercial species. These theories traveled far: from their "experimental sites" in some colonies to the metropolitan areas where future foresters were trained then implemented in yet other colonies. Through the flow of personnel, publications, and reports, new ideas concerning forestry management were shared among colonial foresters. Something of the same process is now happening with respect to the production of sustainable forest management (SFM) discourse; however, now the forum for the exchange of views is represented by the highly efficient machinery of transnational organisations like ITTO. Gale's account of the SFM discourse clearly reveals how diverse players in the world timber market may have a stake in valorizing a single paradigm thus contributing to its establishment and dominance in state policies.

At the most intimate level of analysis, political ecology seeks to examine the effects of environmental change on local peoples. With the advance of the modern global market, the interconnections between disparate local histories are becoming more difficult to disentangle. The effects on forests and their inhabitants are uneven from place to place. The studies in this book reveal problems in romanticizing or castigating the "poor" who bear the brunt of environmental degradation. Poverty, as mentioned earlier, is not in itself a cause of forest destruction. Rather, a political ecology approach would have to examine the play of class and power interests both in "creating" the idea of the poor (Armstrong 1998; Dove 1993; Li 1999) and in using such images to advance external agendas.

Apart from the environmental effects of land and resource expropriation, upland peoples are also vulnerable to the concrete manifestations of power—the psychological and physical threats of coercion, cooptation, rent-capture and so on (de Jong *et al.*, Chapter 4). This is observed at the state level, but also at the village level, as demonstrated by Rhee (Chapter 6) in his discussion on local people's response to decentralization and the prospects of improving their political status and economic resources. The result is, predictably, the growth of conflict and emergence of new alignments of political interests. Case studies of this ilk provide a much-needed corrective to romantic notions of peasants (and forest-dependent communities more generally) common among conservationists. Given local peoples' histories of dispossession and deprivation, there is no reason why they should reject socio-economic benefits long denied to them. The "poor" are enmeshed in wider market structures and they are not necessarily inclined to favor efforts to control resource exploitation. Part of this is also due to the hegemony of state ideologies. To the degree, for example, that Indonesians saw President Suharto as the "father of development" (Abe, Chapter 5) they may have internalized the oppressive structures under which they chafe and of which they are a part.

Among the peasants studied by Ichikawa (Chapter 7), logging has given them increased economic opportunities. Moreover, they are the unseen "supporters" of the timber economy. The rice they plant and the products they trade subsidize *in situ* timber operations—which gives them increased interests in the continuation of the logging industry. They are embedded in the political economy of timber and have economic interests in it. This a more subtle instance of the process of cooptation. As noted earlier, the territorial organization of ethnic groups historically was linked to the wider search for lucrative forest resources. Rhee's discussion (Chapter 6) shows how this old process is recurring in the forest-rich districts in Indonesia today. There the shift from central to district to local appropriation of rich forest resources shapes for an important part the decentralization process and the power and control that is held at the district and local levels.

While forests were utilized and managed by local communities, no irreversible changes occurred. Tropical forests began to deteriorate as modern state formations introduced centralized management regimes, which triggered the loss of local autonomy and control over self-support systems (Guha 1989; Harper 1998; Potter, this volume). However, the globalization of national tropical forests, and its impacts on the forest itself and local livelihoods may take unpredictable directions (Abe, Chapter 5). One example is the exploitation of so-called "minor" or non-timber forest products. This has also affected the forest and people surrounding it, but not in a straightforwardly destructive manner, as is in the case of timber. The vegetation along the Barito river in South Kalimantan seems to be natural but is in fact artificially enriched with rattan and its supporting trees. These rattans, which provide raw materials for weaving *lampit* (rattan mat), are actively cultivated by nearby villagers and sold to processing factories. Rattan is one of the few major sources of cash-income for the riverine villagers (de Jong *et al.*, Chapter 4). This internationalization of Indonesia's tropical forest has changed the condition of the forest, but these changes are hardly comparable to those caused by logging or conversion of peat lands.

Lampit, which was traditionally valued within Borneo only, can also be used in Japanese houses, particularly in the summer months. At the factories, almost all *lampit* (*tomushiro* in Japanese) from the Barito is now exported to Japan. Thus, this riparian forest in Borneo also became directly linked to the Japanese markets, just like logging concessions in East Kalimantan became directly linked to European, Japanese, and Chinese markets. Sales of *lampit* fluctuate according to the Japanese climate, which in turn affects the purchasing price of rattan in Borneo. However, competing producers from China, and especially China's policies affecting its own bamboo sector (Zhong *et al.* 1998), combined with ill-conceived policies in Indonesia, affect Indonesia's volume of rattan exports (Belcher *et al.* 2000; de Jong *et al.* this volume). This, to some extent, led to the collapse of production from rattan gardens in East Kalimantan (Belcher *et al.* 2000) and the voluntary replacement by the owners of diverse, forest-like rattan gardens to monospecific oil palm plantations under PIR arrangements.

This particular example demonstrates the inter-connections among local communities around the world. For, just as Japanese demand for this kind of rattan and Chinese policies affect communities in Borneo so, conversely, as Knight shows (Chapter 10), the rise of Southeast Asian dominance in timber production has helped in the disintegration of timber-growing economies in upland Japan. This economically motivated opening up of Japanese markets for tropical (and, latterly, temperate) woods profoundly affects national timber producers. Not only does the import of *gaizai* (Japanese for "foreign wood") destroy the economic basis that allowed Japanese foresters to subsist, it also fundamentally affects a system of values and a religion-like philosophy of life. There traditional foresters lose what seems to them their identity as timber growers, since the planting of trees reflects the continuity of the household and the family history. Another implication of Knight's study is the failure of planners to think in the long-term. What happens to timber-producing communities when the income opportunities they had embraced so enthusiastically decades back are threatened?

The case of dwindling rattan production in East Kalimantan demonstrates that some of the proposed solutions, like devolution of forest management to local communities, will not detach forest management from global processes. Ironically, in view of the globalization forces that had opened up markets in the first place, regulatory attempts to redress conditions—such as the growing impoverishment of upland peoples—may have to be global in their scope and scale (Thompson, Chapter 8). However, as Gale points out, as long as the current paradigms persist, such efforts will not have a fair chance of working.

An issue that warrants reflection is the particular characteristics of the Asian forests. As discussed above, the tropical forests are not homogenous and nor are their histories. The tropical forests of Southeast Asia, as Abe stresses in Chapter 5, are fundamentally distinguishable by ecological and biological criteria. Seldom is this reflected in political ecology discussions on those forests, even though part of political ecology thinking does consider the ecology of natural resources. But not only do forests differ biologically, some forest with the same biology are viewed differently from different actors, from different regions, while the way forests are looked at may change over time and be given different status. Both the biology of forests and what they are used for, both practically and ideologically, ultimately should determine the level of analysis, as both affect their capacity for regeneration.

Chapters 2 (Potter) and 7 (Ichikawa) show that the idea of "forests" is not without its own problems. Potter's chapter most fully fleshes out the implications of forest diversity on policy, management, and politics. Ecologically, the effect of colonial policies was to substitute homogeneity for the diverse social-ecological and management systems of precolonial forest populations (e.g. Poffenberger 1990; Scott 1998). The enclosure of teak-rich forests provided a ripe ground for on-the-ground conflict between foresters and local communities (see also Peluso 1992d). Forest diversity is exploited in the land management systems of the Iban swiddeners studied by Ichikawa. For these farmers, two main types of agricultural land are available: swamplands and hill forest. Thus, they are able to take advantage of one and/or the other as economic needs change; this enables them to practice a form of resource management in which forests are allowed time to grow back. The Iban's management system also provides a diversity of necessary crops that helps them to tide over periods of scarcity and famine. However, for the British foresters studied by Potter, diversity was a problem that had to be overcome rather than exploited.

The idea of forest—and more broadly of the environment—as shown in the parallel chapters by Abe and Gale is often subjected to discursive manipulation. In Malaysia, for example, the actual extent of land remaining under forest cover is fiercely disputed, with government agencies using a variety of criteria—depending on what the statistic is being used for—to determine what constitutes "forest" (Sahabat Alam Malaysia 2001). These criteria may not correspond to their scientific or indigenous counterparts. Abe argues that the development of peat swamp forests in Indonesia was ultimately a last-ditch attempt by the Suharto regime to assure the populace that the government was still concerned to engender development while Gale argues that the sustainable forest management discourse worked out at ITTO may inhibit a shift towards ecosystem-based forest management. The ideological use of forests has increasingly become a cross-national issue, as the all-important national policy environment is subject to transnational machinations, a point made most fittingly by Gale (Chapter 9).

In view of the historical focus, temporal concerns are critical to the analyses here. The workings of governments and administrators, as Chapters

2 to 4 show, are conditional on how trade relations, power interests, and management bureaucracies developed in the past. In this regard, the question of social relationships discussed earlier—between growers and suppliers, extractors and traders, between states and the labor they have access to—becomes crucial to the analysis. Less considered in most historical analyses is the question of the future, a point taken up by Thompson (Chapter 8). The values of forests should not be limited to contemporary generations but be extended to future generations as well. In current economic theory this incorporation of the needs of future generations is hardly possible, because of the low value that the economic interests of future generations are given in the accounting of economic transactions. This, however, is a result of some of the basic premises in economic theory, which, as Thompson argues, should be revised. For while misleading statistics are used as the basis for government policies, the future is already being jeopardized by the degradation of the tropical forests.

To summarize, the chapters in this volume adopt a variety of approaches as they examine different dimensions of forest politics and economics in their respective topics of study. As befitting the diffuse origins of the political ecology approach, the tendency in this volume is less towards a coherent presentation of theory, than in applying a certain set of assumptions regarding linkages between state, resources, people, time, and space to the various issues in Southeast Asian tropical forests today. The volume reflects, both in content (through the topics of the chapters) and in form (through the different disciplinary orientations of the contributors), a political ecology way of understanding, explaining, and critiquing contemporary resource problems. The strength of political ecology is the possibility of critique, which is the first step in devising appropriate solutions. With regard to solutions, neither a fully local nor a globally imposed paradigm is appropriate. As the effects of degradation proceed apace, a multitude of actions will need to be devised, each one flexible to the problems it addresses and sensitive to the local forest histories upon which it impacts. Given the interlinkages among communities, agents, and states around the world, straightforward solutions will not be forthcoming.

Notes

1 This concern was also an offshoot and in large part continuous with earlier (1960s to 1970s) anxieties over the quality of life under the impact of industrial capitalism. In this earlier wave of environmentalism,

concerns were less focused on the fate of forests than by the so-called "brown issues" of pollution and energy use, which eventually coalesced into the beginnings of the "Green Movement". Although the actions and prescriptions of the Greens are not absent from current deforestation debates, this volume does not specifically address the impact of Euro-American environmentalism on local struggles over the tropical forests. As such, our review of contemporary issues begins from the 1980s onwards when tropical forests and species loss became established as public issues in the region's countries.

Bibliography

Andaya, B., and L.Y. Andaya. 1982. *A History of Malaysia.* London: MacMillan.

Andaya, L. Y. 1975. *The Kingdom of Johor, 1641–1728.* Kuala Lumpur: Oxford University Press.

Anon. Editor. 1992. *Logging against the Natives of Sarawak*, 2nd edition. Kuala Lumpur: INSAN.

Armstrong, R. 1998. Insufficiency and Lack: Between Production and Consumption in a Longhouse Economy 1909–1996. *Journal of the Royal Anthropological Institute* 4:511–530.

Assies, W., van der Haar, G., and A. Hoekema. 1999. *The Challenge of Diversity: Indigenous People and Reform of the State in Latin America.* Amsterdam, Thela Thesis.

Barr, C. 1998. Bob Hasan, the Rise of Apkindo, and the Shifting Dynamics of Control in Indonesia's Timber Sector. *Indonesia* 65 (April).

Belcher, B., *et al.* 2000. Resilience and Evolution in a Managed NTFP System: Evidence from the Rattan Gardens of Kalimantan. Paper presented at the workshop: Intermediate Management Systems Forest Products. Lofoten, Norway, June 2000.

Blaikie, P., and H. Brookfield, eds. 1987. *Land Degradation and Society.* London: Methuen.

Browder, J.O. and B.J. Godfrey. 1997. *Rainforest cities: Urbanization, Development and Globalization of the Brazilian Amazon.* New York: Columbia University Press.

Bryant, R.L. 1992. Political Ecology: An Emerging Research Agenda in Third-World Studies. *Political Geography* 11:12–36.

Bryant, R.L., and S. Bailey. 1997. *Third World Political Ecology*. London: Routledge.

Bunker, S.G. 1985. *Underdeveloping the Amazon: Extraction, Unequal Exchange, and the Failure of the Modern State*. Urbana, IL: University of Illinois Press.

Cant, R.G. 1973. *An Historical Geography of Pahang*. Monograph no. 4. Kuala Lumpur: Malaysian Branch of the Royal Asiatic Society.

Chibnik, M. 1991. Quasi-ethnic Groups in Amazonia. *Ethnology* 30(2): 167–182.

Conklin, B.A., and L.R. Graham. 1995. The Shifting Middle Ground: Amazonian Indians and Eco-Politics. *American Anthropologist* 97:695–710.

Conklin, H.C. 1957. *Hanunóo Agriculture: A Report on an Integral System of Shifting Cultivation in the Philippines*. Forestry Development Paper no. 12. Rome: FAO.

Cooke, F.M. 1999. *The Challenge of Sustainable Forests: Forest Resource Policy in Malaysia, 1970–1995*. St. Leonards, NSW, Australia: Allen & Unwin and Honolulu: University of Hawaii Press.

de Jong, W. 1997. Developing Swidden Agriculture and the Threat of Biodiversity Loss. *Agriculture, Ecosystems and the Environment* 62:187–197.

de Jong, W., Chokkalingam, U., and G.A.D. Perera. 2001. The Evolution of Swidden Fallow Secondary Forests in Asia. *Journal of Tropical Forest Science* 13(4):800–815.

Dauvergne, Peter. 1997. *Shadows in the Forest: Japan and the Politics of Timber in Southeast Asia*. Cambridge, MA: MIT Press.

Dennis, R., A. Hoffmann, G. Applegate, G. von Gemmingen, and K. Kartawinata. 2001. Large-Scale Fire: Creator and Destroyer of Secondary Forests in Western Indonesia. *Journal of Tropical Forest Science* 13(4):786–799.

Dove, M.R. 1983. Theories of Swidden Agriculture and the Political Economy of Ignorance. *Agroforestry Systems* 1:85–99.

——— 1985. Plantation Development in West Kalimantan I. *Borneo Research Bulletin* 17 (2): 95–105.

——— 1993. A Revisionist View of Tropical Deforestation and Development. *Environmental Conservation* 20:17–24.

——— 1996. So Far from Power, So Near the Forest: A Structural Analysis of Gain and Blame in Tropical Forest Development. In C. Padoch and N.L. Peluso, eds., *Borneo in Transition: People Forests, Conservation, and Development*, 41–58. Oxford: Oxford University Press.

Edmund, D. and E. Wollenberg. 2001. Historical Perspectives on Forest Policy Change in Asia: An Introduction. *Environmental History* 6 (2): 190–212.

Glover, D. and T. Jessup, eds. 1999. *Indonesia's Fires and Haze*. Singapore: Institute of Southeast Asian Studies.

Greenberg, J.B., and T.K. Park. 1994. Political Ecology. *Journal of Political Ecology* 1:1–12.

Gudeman, S. 1986. *Economics as Culture: Models and Metaphors of Livelihood*. London: Routledge & Kegan Paul.

Guha, R. 1989. *The Unquiet Woods: Ecological Change and Peasant Resistance in the Himalaya*. Berkeley: University of California Press.

Hardenburg, W.E. 1912. *The Putomayo, the Devil's Paradise*. London: T. Fisher Unwin.

Harper, T.N. 1998. The Orang Asli and the Politics of the Forest in Colonial Malaya. In R.H. Grove, V. Damodaran, and S. Sangwan, eds., *Nature and the Orient: The Environmental History of South and Southeast Asia*, 936–966. Delhi: Oxford University Press.

Hatch, T. 1982. *Shifting Cultivation in Sarawak—A Review*. Technical Report no. 8, Soils Divsion Research Branch, Department of Agriculture, Sarawak.

Hatch, T., and C.P. Lim. 1978. *Shifting Cultivation in Sarawak*. Report of the Workshop on Shifting Cultivation, Kuching, Sarawak.

Hecht, S., and A. Cockburn. 1990. *The Fate of the Forest: Developers, Destroyers and the Fate of the Amazon*. New York: Harper Perennial.

Hirsch, P., and C. Warren. Editors. 1998. *The Politics of Environment in Southeast Asia: Resources and Resistance*. New York: Routledge University Press.

Hong, E. 1987. *Natives of Sarawak*. Penang, Malaysia: Institut Masyarakat.

Hurst, P. 1990. *Rainforest Politics: Ecological Destruction in Southeast Asia*. London: Zed.

Kartasubrata, J. 1993. Indonesia. In: *Sustainable Agriculture and the Environment in the Humid Tropics*, 393–439. Washington, D.C.: National Academy Press.

King, V. 1993. *The Peoples of Borneo*. Oxford: Blackwell.

Leigh, M. 2000. The Political Economy of Tropical Forestry in Sarawak: 1950–1990. Paper presented at the Symposium: The Political Ecology of Tropical Forests in Asia: Historical Perspectives. Japan Center for Area Studies, National Museum of Ethnology. Osaka, November 28–30, 2000.

Li, T.M., ed. 1999. *Transforming the Indonesian Uplands: Marginality, Power and Production*. Amsterdam: Harwood Academic Publishers.

Lim, T.G. 1976. *Origins of a Colonial Economy: Land and Agriculture in Perak 1874–1897*. Penang: Universiti Sains Malaysia.

Lye T.-P. 2000. Forest, Bateks, and Degradation: Environmental Representations in a Changing World. *Tonan Ajia Kenkyu (Southeast Asian Studies)* 38:165–184.

Mayer, J.H. 1996. *Trees versus Trees: Institutional Dynamics of Indigenous Agroforestry and Industrial Timber in West Kalimantan*. Ph.D. Thesis, University of California, Berkeley.

Mintz, S.W. 1986. *Sweetness and Power: The Place of Sugar in Modern History*. New York: Penguin.

Momberg, F., R.K. Puri, and T. Jessup. 1997. Extractivism and Extractive Reserves in the Kayan Mentarang Nature Reserve: Is Gaharu a Sustainable Manageable Resource. In K. Worm and B. Morris, eds., *People and plants of Kayan Mentarang*, 165–180. London: Worldwide Fund for Natura, Indonesia Programme.

Neumann, R.P. 1992. Political Ecology of Wildlife Conservation in the Mt. Meru Area of Northeast Tanzania. *Land Degradation & Rahabilitation* 3: 85–98.

———— 1996. Forest Products Research in Relation to Conservation Policies in Africa. In M. Ruiz Perez and J.E.M Arnold, eds., *Current Issues in NTFP Research*, 161–176. Bogor: Center for International Forestry Research.

Neumann, R., and E. Hirsch. 2000. *Commercialisation of Non-Timber Forest Products: Review and Analysis of Research*. Bogor: Center for International Forestry Research.

Oldfield, M.L., and J.B. Alcorn, eds. 1991. *Biodiversity: Culture, Conservation, and Ecodevelopment*. Boulder, CO: Westview.

Padoch, C. and C.M. Peters. 1993. Managed Forest Gardens in West Kalimantan, Indonesia. In C.S. Potter, *et al*, eds., *Perspectives on Biodiversity: Case Studies of Genetic Resource Conservation and Development*, 167–176. AAAS Washington.

Parnwell, M.J.G., and R.L. Bryant. Editors. 1996. *Environmental Change in South-east Asia: People, Politics And Sustainable Development*. London: Routledge.

Peet, R., and M. Watts, eds. 1996. *Liberation Ecologies: Environment, Development, Social Movements*. London and New York: Routledge.

Peluso, N.L. 1983. *Markets and Merchants: The East Kalimantan Forest*

Product Trade in Historical Perspective. Masters Thesis, Cornell University.
————— 1992a. The Political Ecology of Extraction and Extractive Reserves in East Kalimantan, Indonesia. *Development and Change* 23(4):49–74.
————— 1992b. The Rattan Trade in East Kalimantan. In D.C. Nepstad and S. Schwartzman, eds., Non-Timber Products from Tropical Forests: Evaluation of a Conservation and Development Strategy, [*Advances in Economic Botany* 9: 115–128. New York Botanical Garden, New York].
————— 1992c. The Ironwood Problem: (Mis)management and Development of an Extractive Rainforest Product. *Conservation Biology* 6(2):210–219.
————— 1992d. *Rich Forests, Poor People.* Berkeley: University of California Press.
Poffenberger, M., ed. 1990. *Keepers of the Forest: Land Management Alternatives in Southeast Asia.* Wesport: Kumarin Press.
Potter, L., and J. Lee. 1998. *Tree Planting in Indonesia: Trends, Impacts and Directions.* Bogor: Center for International Forestry Research, Occ-asional Paper; no. 18.
Repetto, R., and M. Gillis, eds. 1988. *Public Policies and the Misuse of Forest Resources.* Cambridge: Cambridge University Press.
Ross, M.L. 1996. *The Political Economy of Boom and Bust Logging in Indonesia, the Philippines, and East Malaysia, 1950–1994.* Princeton: Princeton University Press.
Sahabat Alam Malaysia. 2001. *Malaysian Environment: Alert 2001.* Pulau Pinang: Sahabat Alam Malaysia.
Schmidt-Vogt, D. 2001. Secondary Forests in Swidden Agriculture in the Highlands of Thailand. *Journal of Tropical Forest Science* 13(4). pagination.
Scott, J. 1998. *Seeing Like a State: How Certain Schemes to Improve the Human Condition Have Failed.* New Haven: Yale University Press.
Sellato, B. 2001. *Forest, Resources, and People in Bulungan. Elements for a History of Settlement, Trade, and Social Dynamics in Borneo, 1880–2000.* Center for International Forestry Research, Bogor.
Sunderlin, W. & I.A.P. Resosudarmo. 1996. *Rates and Causes of Deforestation in Indonesia: Towards a Resolution of the Ambig-uities.* Bogor: Center for International Forestry Research, Occasional Paper 9.

Weinstock, J.A. and S. Sunito. 1989. *Review of Swidden Agriculture in Indonesia*. Directorate General of Forest Utilization, Forestry Studies, Field Document No: II–1, Jakarta.

Wilson, E. O., ed. 1988. *Biodiversity*. Washington, D.C.: National Academy.

Wolf, E.R. 1982. *Europe and the People Without History*. Berkeley: University of California Press.

Zahid E. 1990. "The Orang Asli Regrouping Scheme: Converting Swiddeners to Commercial Farmers." In V.T. King and M.J.G. Parnwell, ed., *Margins and Minorities: The Peripheral Areas and Peoples of Malaysia*, 94–109. Hull, UK: Hull University Press.

Zhong, M., M. Fu, B. Belcher and M. Ruiz-Perez. 1998. Effects of Social Economy and Policies on Production Management Systems—A Case Study of China's Bamboo Industries. *Forestry Economics* 3(1): 22–34.

2
Forests versus Agriculture: Colonial Forest Services, Environmental Ideas and the Regulation of Land-use Change in Southeast Asia

Lesley Potter

This paper compares the activities of a number of colonial forest services (or their precursors) in Southeast Asia from about 1875 to 1921, examining particularly the relative importance accorded to forests and agriculture and the role of environmental ideas in regulating forest conversion. The colonial states and their respective forestry departments were evolving throughout this period. As such, this paper's initial comparisons start with three regimes, expanding in later sections to include others. The aim is to examine both changes and continuities and to seek answers to a number of specific questions. As emphasis shifted from slow-growing hardwood species, like teak (*Tectona grandis*), to the more diverse dipterocarp and "wildwood" forests, what was the impact on ideas about the "value" of the forests? Specifically, how successful, as competitors for land, were forestry agencies against the demands of other users, especially commercial agriculture, and what was believed to be forests' environmental importance? To what extent was it possible to transfer management techniques developed in one kind of forest to another? To what extent were colonial foresters interested in and aware of what their counterparts were doing under different regimes? Finally, to what extent were environmental ideas invoked for political purposes, as devices to control or restrict indigenous activities in the forests or to legitimate the plans of forest departments?

A study of the historical interaction between politics and ecology, in a "politico-ecological history" of Southeast Asian forests, must also include economics. This is the important third axis in an essentially triangular relationship. Colonial regimes, seeking to reap maximum benefits from their territories, often anticipated those as coming from cash crops rather than forests, except for a few notable timber species or non-timber products with a guaranteed market. The rationale for land use change, and particularly forest conversion, was often expressed in direct economic terms. Despite the ecological "value" sometimes ascribed to forests, economic considerations would usually guide political decisions. Forestry departments might disagree with those decisions, but final outcomes would likely depend on their relative importance in the colonial hierarchy and their willingness to "lobby" for particular forests.

Demography also had an impact, though an uneven one. The total size of Southeast Asia's population increased markedly over the period, producing increased pressures for forest conversion in some territories. For example, between 1880 and 1920 the population of Java almost doubled from roughly 20 million to 34 million (Gooszen 1999). By the second decade of the 20th century, agricultural frontiers were closing in the Javanese uplands and the Burmese delta, yet other areas remained thinly peopled and the pressures on their forests correspondingly light. It was in the heavily populated areas that the debate was fiercest regarding the respective roles of forests and agriculture. A contrary situation obtained, however, in the lightly populated Federated Malay States. There the exotic rubber *Hevea brasiliensis*, a commercial tree crop, was encouraged to make considerable inroads into forested land once its market was assured. In such a situation any protests by foresters would be over-ruled by stronger interests in the colonial polity. In most Southeast Asian colonies it was considered inevitable that a large part of the forests would eventually be converted to some form of agricultural use. The only exception was the Philippines where, continuing the precedent set by its Spanish predecessor, the Bureau of Forestry exerted greater power.

The General Context of State Forest Management in Asia

The orthodox "European" scientific forestry paradigm evolved in Germany with the concept of the *standard tree* that would produce a high and calculable wood yield (Lowood 1990). The aim was to transform existing forests into assemblages of standard trees, preferably of the most valuable types, with predictable timber crops. These ideas were codified into a training program with the establishment of forestry schools in parts of Germany and France by 1820. Such a program began to be exported to

Europe's overseas colonies in the 1850s. Burma (then part of British India) and Dutch Java were particularly susceptible to such ideas because of their valuable teak forests, seen as essential for European shipbuilding. In Java the teak forests already resembled the plantations that were the natural end point of the German system. However, most forests in the rest of Southeast Asia were very different, constituting one of the world's great reservoirs of biodiversity through which "marketable" trees were often thinly scattered. It is ironic that scientific forestry was itself being challenged in Germany and France, in the "back to nature" movement, while this export of the model was taking place (Ciancio and Nocentini 2000).

The icon of colonial forest departments was the forest *reserve*, which was created to ensure that a particularly valuable species could be protected and encouraged to regenerate in its natural habitat. It was considered to need protection both from local populations and the demands of competing interests. Once the trees were grown a scientific silvicultural system would be worked out for their management. This was mainly a form of "selective logging." Teak trees reaching marketable size would be ring-barked or "girdled" and allowed to stand in the forest to dry for two years before cutting. In the reserves the aim was to manipulate natural conditions to ensure an increased proportion of the valuable species in the stand. The assumption was that only the trained forester could understand the forests: non-foresters were perceived as generally hostile, intent on forest destruction.

The diversity of Southeast Asian forests was exploited by local peoples, both as the basis for trade in plant and animal products and the satisfaction of a variety of needs. The "plantation" model of scientific forestry and the creation and management of reserves implied an unacceptable level of interference with local cultures. Forest departments were universally disliked by local populations and the control of the forests contested. Other civil servants, especially those dealing with land or settlement issues, would sometimes support the people in these contests. The stage was thus set for a series of confrontations, which revealed much about local power structures and struggles over resources. Such topics lie at the heart of political ecology (Bryant 1992).

To give added legitimacy to their activities, foresters drew on environmental ideas that had been honed through expatriate experiences in India and various island colonies like Mauritius (Grove 1995). The ideas related particularly to the impact of forests on climate, both temperature and rainfall (the *desiccation* theory), and hydrology. Forests were seen to be moderators of stream flow (the *sponge* theory), thus reducing the likelihood

Photo 2-1. *Baccaurea griffithii* (Malay *buah tampoi*), a common forest fruit. The diversity of Southeast Asian forests was exploited by local peoples, both as the basis for trade and the satisfaction of a variety of needs. *Photo by Lye Tuck-Po*

of flood and drought. Where the trees themselves were considered valuable, as yielding a commodity in demand (such as gutta percha [a resin of the family Sapotaceae, including various species of the genera *Palaquium* and *Payena*], camphor [*Dryobalanops* sp., especially *D. sumatrensis*] or jelutong [*Dyera* spp.]), their physical destruction might be lamented in ecological as well as economic terms, the ecological arguments being used to apply moral pressure to the fellers and to bring about changes in policy.

Forests and the trees they contained possessed different meanings depending on the context. Trees felled to make way for commercial agriculture would be viewed by colonial authorities as either largely worthless or simply standing in the way of progress. Swidden farmers burning small forest patches were often castigated as "robber farmers" (with greater outrage from observers if the trees happened to be marketable species). Rubber or tobacco growers burning entire forests for their estates would be hailed as noble pioneers. Double standards were rife.

Articulating the peoples' views, to compare with those of administrators or scientists, has been an important focus to recent work in political ecology (see, for example, papers in Stott and Sullivan 2000). In historical context it is more difficult to hear such "subaltern" voices directly, although acts of protest are usually well-documented. These may include violating or ignoring forest "rules," transgressing forest boundaries, theft and arson (Peluso 1992; Bryant 1997; Peluso and Vandergeest *in press*). While some effort will be made here to document local opinion, this is not the main focus of this paper.

Forests growing under a strongly seasonal rainfall regime tend to be more open and less diverse than tropical moist forests, where dry seasons are less regular. In Southeast Asia the deciduous teak typifies the former; the various species of the family Dipterocarpaceae categorize the latter. Relatively slow-growing trees produce dense, high quality timbers. If at the same time they can be used for shipbuilding, like teak and the Philippine molave (*Vitex parviflora*)[1], their value in the 19th century was greatly enhanced. The dipterocarp forests lacked these special qualities and were much less sought after. Apart from some early commercial export development in the Philippines and smaller-scale efforts from British North Borneo, they were used primarily for fuelwood and cheap construction timber.

The following discussion will be divided into three chronological sections:

(1) *The position in 1875*: 1875 is regarded as a kind of watershed for the older forest departments (Burma, Java and the Philippines), for by then basic management systems were already in place;

(2) *The period 1875–1900*: this was an era of rapid and fundamental change, with movements toward formal forest organization being extended to other territories, such as parts of the Malayan Peninsula (now Peninsular Malaysia) and Singapore;

(3) *The period 1901–1921*: interest shifts to the dipterocarp forests, as American commercialism in the Philippines introduces a different approach and the expansion of *Hevea* rubber brings new challenges.

The Position in 1875:
Lower Burma, Java and the Spanish Philippines

Lower Burma

The forests of Lower Burma came under British control after two wars in 1824 and 1852. These forests contained scattered teak trees that the British required for ship-building and, later, the Indian railways. Teak represented ten per cent or less of the forest species in the uplands of Pegu and

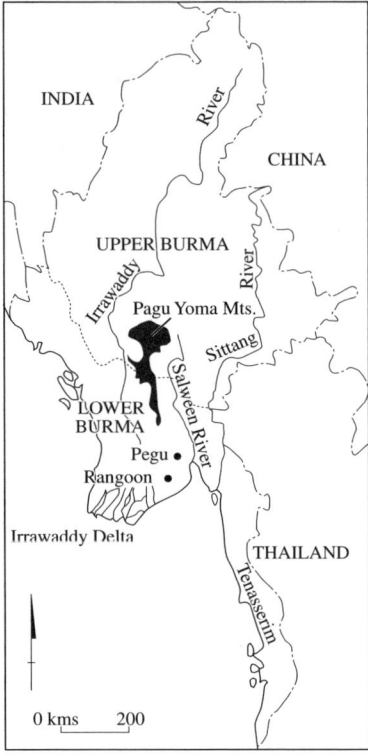

Figure 2-1. Burma

Tenasserim (see Figure 2-2), which were populated by minority Karen people. The year 1875, ten years after the Burma Forest Rules first defined "Government Forests" in which *reserves* could be demarcated, represents the serious beginning of that system in Southeast Asia. The idea was to be copied in other territories, with everywhere an impact on local populations through prohibitions on cutting, collecting, and grazing. Some reserves, intended to protect the best teak areas, incorporated *taungya*[2] plantings of teak by local Karens. The Karen were encouraged to plant and tend teak trees along with their subsistence crops in exchange for retaining rights to reserve land. The reserve system was still in its infancy in 1875, and was confined to the upland teak areas, especially in Pegu. Reserve formation was accompanied by demarcation of boundaries on topographical and ethnic grounds.

When the work of selecting reserves was taken in hand vigorously in 1875–1876 it was found essential to demarcate *an outer line* which should exclude Burmese or hamlets fringing the plains, and also to fix *an inner line* practically marking off Taungya lands within which the Karen tribesmen might continue to exercise their distinctive method of hill cultivation (Nisbet 1901b: 61). The Karen later responded that they felt "like pigs in a pen" as a result of these boundaries (*Resolution* 1881–1882). The demarcation of boundaries in the forests demonstrated the power of the Forestry Department to control land use, especially indigenous agriculture, which was subordinated to timber production. Karen cultivation was tolerated wherever it helped to regenerate teak: however, it was commercial rice cultivation on the plains, and more especially in the Irrawaddy delta, which was seen to have value.[3]

After 1870, there was a rapid increase in commercial rice cultivation in Lower Burma.[4] In that context, it is useful to read Kurz's report on the forests of Pegu (1875), especially his discussion of the Irrawaddy Delta forests. Kurz, a botanist from the Calcutta Botanic Garden, was influenced by the "desiccationist" theories in vogue there,[5] although he did question some aspects of these ideas. After quoting a speech by a fellow botanist about the relationship between climate and forests,[6] Kurz argued that there would be no adverse climatic effects "even if all the forests on alluvial ground are removed, provided that those on the hills are maintained" (Kurz 1875: 67). Government officers, intent on encouraging rice production in the Delta, would have welcomed such a conclusion.

Kurz was, however, of the opinion that deciduous trees, such as teak, did little for the climate: it was evergreen trees that counted. That idea would not have been popular with the forestry establishment, nor his suggestion that the Karen plant evergreen tree crops and bamboos (though he was critical of the amount of teak wood that they consumed).[7] In considering the use of the delta forests, Kurz was careful to distinguish between the different assemblages and the relative benefits of cultivation or conservation. The "littoral forests" lining the streams in the delta could be easily converted to rice lands, but contained some good trees such as *penlay kanazu* (*Heritiera* sp., types of mangrove). With some prescience, Kurz suggested that fuel reserves might be made there in the areas least suited for rice growing. The coastal mangroves, he felt "discharged some important functions" and should remain, as should the seasonally flooded forests on deep alluvium. Most other forests he considered of little value, their soils more suitable for agriculture. There was a scattering of (inferior) teak trees through some of them, the worth of which would be felt only as the area of agrarian lands increased (p. 71). While Kurz could not have foreseen the scale of development which would take place on the deltaic lands, his careful analysis gives lie to the claim that those forests were seen simply as wasteland (Adas 1983: 101).[8]

Burma, being part of colonial India at the time, had the comparative experience of the various Indian states to draw upon, although its situation was different in many respects. When the monthly journal the *Indian Forester* was started in 1875, it provided a vehicle for exchange of views both across "British India" and gradually throughout the region. While this periodical has been belittled by contemporary writers as "the flagship journal of scientific forestry" (Sivaramakrishnan 1996), it was more than that. Especially in the later years of the 19th century and the first decade of the 20th, it was the vehicle for diverse exchanges of views regarding forest policy, many of them critical of current practice, and it began to incorporate considerable amounts of comparative material. Accordingly, the journal widened the knowledge base and also revealed the complexities of the forest situation under different colonial jurisdictions.

Java

In Java foresters were primarily interested in teak. However, the position was different, in that the teak areas (a block in Central and East Java; see Figure 2-2) already resembled plantations, with a high percentage of the valued trees in the forest. Java too published its first Forest Regulation in 1865 following the reestablishment of its forest service,[9] at the same time abolishing the planting of teak by forced labor. The teak forests were divided into those under regular management and those for local use, while outside the teak areas were unmanaged "wildwood forests" (Boomgaard 1994: 125). In the managed teak areas, new plantations were laid out under guidance of forest personnel, with felling and replanting being done by private tender. Where farmland was limited, the Regulation permitted cultivation of field crops between the rows of young trees, provided the teak was not damaged. This is the first mention of the *tumpangsari* system,[10] which resembled the Karen *taungya* but evolved independently and for a different reason. In 1873 forester Buurman introduced this method successfully in Pekalongan district, which had good soils and a large population. It was only when he was able to expand it on a larger scale after 1881, that it became more widely known and imitated (Becking 1928: 47–48, 53),[11] The reason for its success appeared to be the intensity of the cultivation: "the cultivation of produce of the fields...for a few years

Figure 2-2. Java

between the young teak plants, is very important...because the tillage of the soil that is coupled with it, promotes the growth of the teak" (Cordes 1881: 271). Such intensive cultivation also kept down the growth of grasses such as *Imperata cylindrica*, one of the main scourges of teak regeneration in Java.

It was only in 1874 that official attention was given to the uplands. Writers such as Junghuhn had written in the 1850s of the decline of forests in upland Java, citing as causes the exploitation for fuelwood and conversion into coffee plantations or settlements (Junghuhn 1850–54: 1, 409–410). Hefner (1990: 41) has described the Government coffee monopoly as dominating the mountain environment, "transforming previously unworked 'wastelands' into lucrative coffee stands... extending from 600 to 1400 m in altitude." The abolition of the Cultivation System in 1870 and the availability of long leases for uncultivated land, led to an influx of European plantations. The (1874) Ordinance on the Alienation of Domain Land required that permission for opening "wasteland" be secured from district level officials rather than village chiefs. This was an attempt to restrict villagers' access to forests and practice of shifting cultivation.[12] At the same time, the relaxation of Cultivation System restrictions on migration resulted in a rush of plains people to the uplands, especially between 1875 and 1885 (Palte 1989: 43). Disease brought a decline in government coffee after 1878, releasing both land and labor.

Against this background of indigenous in-migration, land-use competition, and changing commercial possibilities, environmental considerations were invoked in the uplands. The second Forest Regulation (1874) specifically recognized the value of natural or "wildwood" forests. Based on strongly emerging views from Germany and Switzerland, the regulation stressed the climatological and hydrological values of montane forests. Floods, droughts and interruption of lowland irrigation systems were predicted to result from forest conversion (Lugt 1933: 112–113). With no supporting evidence, the entire blame for upland forest loss was attributed to smallholder shifting cultivation and there were direct efforts to eliminate it, both through the clearing regulations and by the establishment of a force of Forest Police. It is suspicious that the wildwood forests and the shifting cultivators who sought a living in them were suddenly singled out for discussion at a time when European capitalists were being invited to take up "wasteland" on long lease. Given the rush into the uplands and the continued existence of an open agricultural frontier, such a regulation had little chance of success.

Van Gorkum (1874) was critical of the supposed forest/climate links, especially the statement that a reduction in rainfall would result from deforestation. He found no evidence of that, but was more accepting of the "sponge theory" for water control. In British India when the peaks of the mountains were cleared, the mountain streams were affected. Van Gorkum therefore suggested that in Java cultivation should not go to the very top of the mountains, which should be kept under trees or reforested.

The situation was thus different from that in Burma, where the agricultural land was predominantly low-lying and offered no direct competition to the hilly teak forests. In the hills of Lower Burma, the formation of forest reserves provoked opposition from Karen agriculturalists, but they were marginal compared to the majority of the Burmese population. In Java, though the teak forests again formed a separate (mainly limestone) block of flat to rolling land, the true uplands, being rich and volcanic, were intrinsically attractive for both plantations and smallholders. They were more likely to be sites of conflict between agricultural and forest interests, especially as population pressure increased. There was therefore more pressure placed on swidden farmers in Java than in Burma, where hilly land was not much in demand. Upland forest reserves were eventually also formed in Java but they were comparatively few (Boomgaard 1994: 128–129).

The Philippines

In the Philippines the situation was different again, with an initially very small forestry establishment expected to handle the scattered forests of many islands. The Philippine Forest Inspection service was set up in 1863 with one engineer and four aides,[13] following concerns expressed about the sustainability of wood supplies for Manila's shipbuilding yards. First attempts to license woodcutting, limiting free exploitation to people's

Figure 2-3. The Philippines

own needs, evoked much resistance. The area of forests was perceived as so large that restrictions were unnecessary. The complete lack of state organization in the forests in 1865 was compared to that in Java under the VOC[14] (Jordana y Morena 1891: 232). In 1873 more support was given to the Philippine service with an increase of staff to 26, including three engineers, eight aides and 12 guards. It was at that point that its work really began (Jordana y Morena 1891).

While the Philippines had no natural supplies of teak, it was rich in slow growing, salt water resistant woods that could be used for ship-building, especially molave (*Vitex parviflora*), a species of the teak family. Blanco (1837) described molave "a wood so precious, and so much sought after" as growing "everywhere";[15] 40 years later it was noted that "Trees of large dimensions are getting scarce in places where they may be easily removed" (Vidal y Soler, D. 1877: 51). One of the first activities of the Forest Inspection Service was to compile information on the extent of the forest estate for the different islands. While parts of southern Luzon had less than 40 per cent forest, most disturbing were Cebu and Bohol, with only 6.6 per cent and 11.5 per cent respectively of the land remaining timbered (Jordana y Morena 1876).

Engineer Vidal complained that the "former magnificent forests" of Cebu were in a "lamentable state" (Vidal y Soler, S. 1874: 12). Expansion of agriculture, especially the export sugar industry, and increased demands for timber after 1860 have been suggested as the main reasons for this decline (Fenner 1985).[16] Cebu exported lumber up to the early 1870s; cancellation of tree-felling licenses simply sent this trade underground. It was predicted that Cebu's remaining forests could be saved if only the fraudulent trade could be stopped (Jordana y Morena 1876: 105).

Vidal, however, blamed the forest destruction mainly on *cainges* or shifting cultivators. He again invoked the climatic connection: "So much of the humidity depends on the great mass of trees...On the islands where much of this has diminished, one can note an unfavorable change in the climate, as, for example, in that of Cebu." (Vidal y Soler, S. 1874: 36). His summary of the overall situation was terse: "The Philippine forests can supply all the needs of local consumption; if use is not regulated however, sooner or later they will be ruined and the country will be barren and uninhabitable. The forests have to be conserved, with clearing and cultivation allowed only after careful planning" (*Ibid*: 42).

The Forest Regulation of 1873 classified state forests into two groups: a) those where the soil was suitable for agriculture on a permanent basis and b) those which should remain in forest, because of their influence on the climate, the hygienic conditions, or the hydrology of the country. Like its Javanese counterpart of 1874, the Regulation also contained a clause proscribing shifting cultivation or *caingin* (Jordana y Morena 1891; Nano 1951). In attempting to institute scientific management of the forests, the Inspection Service took on a heavy task of exploration, planning detailed study of each forest to calculate how much felling could be allowed. While good on paper, such plans were beyond the capacity of the small forest service to implement. Nevertheless, the Philippines was divided into regions and guards were dispatched to prevent illegal forest use in areas where the timber trade flourished. Having taken on the task of controlling land use in the Philippines, the Spanish foresters would find that to absorb most of their energies in the following 25 years. No reserves were contemplated, simply because of a lack of basic silvicultural information. Villagers remained unfettered in their use of timber and other forest products, unless those were for sale, when a license was required.

From 1875 to 1900:
Burma, Java, the Philippines, and Others

Burma
The passage of the new Burma Forest Rules (1882) followed the Indian Forest Act of 1878, but adapted it to local conditions. A major difference was that Burma had no category of "preserved forest," which under the Indian rules lay between the reserves and the open forest in terms of accessibility. Another category, "village forest," was written in the Rules but never implemented. Following a further war in 1885–1886, Upper Burma came under British administration. Its immense teak forests, worked on long term leases by large (mainly British) private firms, brought into question the policies of encouraging teak *taungyas* as a regeneration technique. Reserves continued to be demarcated, however, and remained a key part of forest policy. Nisbet (1901), reviewing the previous 20 years, had a number of pertinent observations. The rules for creating reserves had been carefully framed so that the Forest Settlement Officer examined rights and claims over the area, excluded land for shifting cultivation and adjudicated on rights of way, forest products, watercourses, and grazing. Six months were allowed for objections before the reserve was gazetted, with the proviso that it could be de-reserved again during the following five years (Nisbet 1901b: 69).[17] He concluded: "In a thickly forested but thinly populated country like Burma it seems especially desirable that forest reservation should, in the interests of water storage and of agriculture generally, be carried out as rapidly and extensively as is practicable, so that fresh cultivation (except among the true hill men) should break ground chiefly on the plains, leaving the cultivable portions of the true forest tracts to be settled only when rendered necessary by the increase of population" (Nisbet 1901b: 70).

On the question of land rights, one of the contentious issues with such reservations, Nisbet made the following comment: "The State, as the inheritor of the rights of the Kingship of Burma, could not but be regarded as having the ownership of all waste land and forest not alienated by grant or under the customary tenure for agricultural occupation. On the other hand, however, the people generally had been, from immemorial time, in the habit of making unrestricted use of the forests for felling wood and bamboos, extracting wood-oil for torches, cutting grass for thatching, grazing their cattle and clearing forest land for temporary or permanent cultivation. Thus, although they had no actual *proprietary rights*, the

Photo 2-2. Teak *taungya* plantations in Burma. Source: "Teak taungya plantations in the Henzada-Maubin division," by C.W. Allen (*Indian Forester* vol. XLII, pp. 533–537, 1916)

people living within the forests and in their vicinity had been accustomed to *privileges* amounting practically to *rights of user*, which required careful consideration in connection with the selection of reserved forests" (Nisbet 1901b: 59–60).

Nisbet could foresee a time when all non-reserved land would be cleared for cultivation, and expressed concern about the rates of forest clearing in "the more populous southern districts," noting that such clearance had changed the attitudes of many civil servants: "Men who formerly scoffed at the departmental efforts made for the formation of forest reserves, after seeing the swiftness with which total clearance of forest growth can be accomplished and the evils following as its result, have personally recently urged that the last remnants of the forests of the lower delta should be preserved for timber and fuel production"(Nisbet 1901a: 249).

The reserves were to remain closed to almost all activities, especially exploitation of timber: "The policy adopted is to confine the work of extraction as far as possible to the unreserved forests, so as thus to utilize the existing supplies of timber of all sorts before these areas are ultimately cleared for permanent cultivation." Up to the end of 1899–1900, reserves covered ten per cent of the total area[18]: Nisbet predicted that eventually they would occupy 12 per cent (Nisbet 1901b: 62). "Outside these reserves enormous tracts of tree-forest and jungle still remain for clearing and cultivation, reservation being for the most part confined to forest land unsuitable for permanent self-sustaining cultivation" (Nisbet 1901a: 249). He also remarked that although the area under rice cultivation had more than doubled in the last 20 years, there was still plenty of forested land to allow for further expansion (Nisbet 1901a: 434).

It is little wonder that the period between 1875 and 1900 saw the clearing of most of the deltaic forests. There was an increase of 1.2 million ha in the area under rice during that time and of these 243,000 ha were established in the last five years of the century alone (Owen 1971). There was a rush of migrants (many from Upper Burma after 1886) to the Irrawaddy delta and the lower Salween and Sittang. With cheap "occupancy rights" as security for loans to cover their initial costs, migrants would find a piece of land and clear it from forest "girdling or drowning out the multi-rooted mangrove trees, clearing underbrush and burning...stumps" (Cady 1958: 158). While a Land Records Department was set up in 1900, it was 1906 before a Department of Agriculture was established.

While Burma's small population and large land area, especially its prospects for rice cultivation, may have minimized the initial impact of the reserves, as Nisbet suggests, this was not the case in India, where the 1878 Forest Act was having an adverse effect on rural society (Haeuber 1993: 56). J. A. Voelcker, an agriculturalist, investigated the situation and suggested a modification of the Forest Act to support agricultural production. This occurred in 1894, the legislation being termed the Voelcker Forest Policy Resolution, which was to apply to the whole of British India (including Burma). The most significant changes were a revised list of forest categories, with more attention to local demands for land and forest resources. The categories were:
- Forests whose preservation is essential on climatic or physical grounds.
- Forests which afford a supply of valuable timber for commercial purposes.
- Minor forests (or village forests) to be managed in the interests of the local population.
- Pasture lands, which may not be forests at all, but whose declaration as "reserved forest" is advisable to obtain settlement of rights ('Op' 1911).

Contemporary observers of the Indian forest scene such as Haeuber (1993) and Jewitt (1995) are skeptical that this change had much effect on what was a quite severe exclusion of local peoples from forest resources. There was initially little discussion in Burma of Voelcker's forest categories: this would come in the following decade, once the removal of the deltaic forests began to impact on the population.

Java

As in India, Javanese were feeling the exclusion from previously available forest resources, as the rules of 1874 had forced villagers in the teak areas to pay for firewood and other products. Peluso (1992) and Boomgaard (1994) document the rise of opposition by the end of the century, particularly as demonstrated by the Samin movement of passive resistance. The *tumpangsari* system, while providing a convenient and cheap means of regenerating teak, also gave poor farmers only a temporary and migratory access to land. With no permanent tenure in the teak forest areas, they became what Peluso has termed "a new kind of forest-dependent rural proletariat" (Peluso 1992: 64). Such groups still exist in Java's teak forests, and they remain disadvantaged (author's fieldwork 1998).

Movement into the mountains continued unabated, provoking further discussion of "Climate forests" by foresters assigned to those regions (Ham 1895). Ham, like van Gorkum before him, was skeptical about the impact of such forests on the local climate and attempted to measure it. He also emphasized the difficulty of maintaining the mountain forests with so many competing land uses, including coffee and other plantations and shifting cultivation, which would often result in grassland through the burning activities of its practitioners. He was against a scattered population, preferring people to farm more intensively and live together in larger villages. He and a few colleagues tried revegetation with leguminous trees, such as *Albizzia molucanna (sengon laut)*. They focused particularly on the "bald" peaks of Central Java, such as Sendoro and Sumbing (Koorders 1895; Tobi 1894). Holle tried for 20 years, without much success, to persuade small farmers to terrace their sloping fields. He was constantly campaigning against shifting cultivation, or "robber farming" (*roofbouw*) (Holle 1892; 1894). As land availability declined by 1900, some were more ready to listen. In an effort to control lowland flooding, the government returned a large portion of its coffee stands to natural forest and relocated their native cultivators downslope (Hefner 1990: 47).

Despite these scattered attempts to improve ecological conditions in the mountains, most foresters were interested only in teak. The Forest Regulations of 1897 divided the teak region into forest *ranges*, to be operated by the Government Forest Service under detailed working plans,

Photo 2-3. Assembling teak logs, Java (note use of oxen). Source: *Het Boschbeheer in Nederlandsch Indie* (Ch. S. Lugt, 3rd printing. Haarlem, Willink, 1933).

and forest *districts*, where plans remained rudimentary and private exploitation continued (Boomgaard 1994: 130). The aim was to bring all teak land under government management, a task entrusted to a special "Planning Brigade." Under the 1897 Regulations the non-teak forests were also divided, into those to be preserved for protection of "climatological and hydrological values," and others. The latter were to provide fuel for railways and sugar industries or be converted into plantations, a process already well under way. Very little attention was given to the "climate forests."

The Philippines

The 1873 Regulation defined the right of the Inspection of Forests to classify lands and then sell those which could be released to private buyers for cultivation. "The speed at which the documents were processed and the low costs to the bidder were powerful reasons for the natives to want to buy the State lands and the demand grew to such proportions that it was impossible...to keep up with the work" (Jordana y Morena 1891: 246). By the end of 1881, more than 2000 applications had been received for "state wastelands" and 100,000 for legitimization of already cultivated land: "The sale of public waste land brought advantage to individuals as it made the ownership of rural property legitimate" (258). As this land classification role was too much work for the staff available, the establishment was increased to 156, with the "office of sales" and "division of public lands" being given to a special commission. Four forest districts were formed: North, Central and South Luzon, together with one covering the Visayas and Mindanao (260).

The Forest Inspection Service had thus accepted the dual role of a joint forest and lands department. It initially controlled all lands except those to be placed under communal village tenure, then released for a small fee those already cultivated, or with the potential to be brought under cultivation. One difficult task was the demarcation of communal lands, supposed to be village-controlled forests where products for local consumption could be obtained freely (Jordana y Morena 1891: 242). However, there were disputes over this demarcation and the Regulations were slow to be implemented. There was strong opposition to the Inspection Service and the timber traders complained constantly. Jordana y Morena (1891: 245) wrote that the Inspection functioned in a "suffocating environment," (*una atmosfera asfixiante*) and that it was a miracle it did not fold.

The service was supported by the monthly journal *Revista de Montes*, which began publication at the Spanish forestry training school in 1877. While reports of activities and personnel involved in the Philippines were carried regularly, most of its focus was directed to Spain and Europe, with some material on the United States. With the exception of occasional articles on Cochin China derived from its French counterpart (*Revue des Eaux et Forets*), there was nothing on Asia. Although they had indirect access to information about plant taxonomy in Southeast Asia, from the botanical gardens at Kew, Paris, and Leiden, the foresters in the Spanish Philippines felt cut off, especially from first-hand exposure to policy development. In 1878 the Head of the Inspection Service asked permission to visit Java and British India, but this was refused by the Spanish authorities. Jordana wrote of the difficulties caused by "The lack of frequent and direct communication between those foreign possessions and our Filipino archipelago," so that his sources were very limited. It may have been that isolation which brought about the development of such a different system of forest management in the Philippines. In his *A Comparative Study of Forests in Philippines, India and Java* (1891), Jordana stated that he wanted to improve the situation in the Philippines along Indian lines. He never had the chance to try that experiment.

If Burma was convulsed by the Third Anglo-Burmese War of 1885, the Philippines was more devastated at the end of the 1890s, experiencing in quick succession a nationalist victory against Spain, a short-lived period of independence, then occupation by the United States. From 1898–1901 a United States military government ruled, during which information was gathered concerning the forest administration of the previous regime. An attempt was made to re-hire former (Filipino) forest guards, whose local knowledge was highly valued. The unique combination of forest and lands activities which had characterized the Spanish service was continued by the American administration. Section 18 of the Philippine Bill of July 1, 1902 gave the Bureau of Forestry the exclusive authority "to examine, classify, delimit and finally determine timberlands and lands more valuable for agriculture, which may be released and certified alienable and disposable and placed under the administrative jurisdiction of the Bureau of Land" (Nano 1951: 50). Forester Ahern praised his Spanish counterparts for their interest in conservation, but "faulted them for failing to do much practical work beyond collecting fees for timber" (Roth 1983: 41).

Other Forest Services and Their Precursors, 1875–1900[19]

Here we will examine the position in the Straits Settlements, and in the territories which in 1896 became the Federated Malay States. We move away from a concentration on high value timbers and relatively simple classifications to the dipterocarp forests, with their enormous complexity and biodiversity. In these forests non-wood products tended to be of greater economic importance than timbers, especially rattans, resins and wild rubbers, such as gutta percha (*Palaquium* and *Payena* spp.). The last commodity, which experienced a boom in the final two decades of the 19th century, was collected from the forests by indigenous people, especially in Borneo, Sumatra and parts of the Malayan Peninsula. The mature trees were generally felled in the process, bringing universal disapprobation from concerned authorities, who predicted the disappearance of the resource. This resin was used to coat under-sea cables, so maintaining a regular supply of high quality material was vital. Much effort was expended in attempting to organize and regulate the gutta percha trade, even though formal forest departments were not yet established in most of the producing territories (Potter 1997).

Straits Settlements[20] *and Federated Malay States*[21]

An item appeared in the *Indian Forester* in October 1885, noting the formation in July 1883 of a small Forest Department for the Straits Settlements, by the Superintendent of the Singapore Botanic Gardens, Nathaniel Cantley. The justification for such a department was said to be a consequence of the deforestation caused by Chinese pepper and gambier or tapioca planters, the former operating mainly in Singapore and the latter in Malacca. It was noted that: "Extensive grass wastes...are to be seen

Figure 2-4. The Malayan Peninsula

throughout the settlements. Our timber supply has fallen short of demand and the climate of the colony is becoming sensibly affected." Cantley expanded on the climatic problems of the lack of tree cover, presumably referring to Singapore:

> The total rainfall has not decreased, but owing to the removal of the tree covering—that great equalizer of rainfall—showers have become less frequent and more local than formerly; and droughts of unprecedented length have occurred, thereby increasing the possibility of epidemics...The hill streams run with greater irregularity and many of the smaller streams have become entirely dried up (Cantley 1883: 489).[22]

It was estimated that Singapore used 810,000 cubic feet of timber per year, more than half in the form of stakes by pepper planters, while firewood (101,000 cubic feet) was required for gambier production and shipping.[23] A number of measures were recommended for the better management of the forests, including setting up reserves for the supply of fuel and small building wood and establishing a body of Forest Police.

Cantley had been trained at the Royal Botanic Gardens, Kew and the role of Kew was to be strong in the subsequent developments of both forestry and "economic botany" in the Straits Settlements and Federated Malay States.[24] After Cantley's death in 1888, Henry Ridley, a botanist, was recommended by Kew and appointed as his successor. He quickly established experiments with the Para rubber (*Hevea brasiliensis*) plants in the Singapore Garden's collection, as he sought an economic method of tapping the trees and attempted to interest planters in the crop. Five years later the activities of the Forest Department came under scrutiny from a cost-cutting Government.

On October 29, 1893, Kew Gardens warned the Colonial Office that the future of the Straits Settlements' Forest Department was precarious.[25] The following week the Officer Administering the Government of the Straits Settlements suggested that the time had come to dispense with further expenditure on the Forest Department, "most of the duties of which can probably be handed over to the Land Office and District Officers" (*Singapore Free Press*, 7 November 1893). Kew protested strongly about the proposed involvement of the Land Office, arguing that "All experience shows that a Land Office in a Colony is supremely indifferent to the fate of its forests while District officers have often proved actually hostile to them" (*Thisleton Dyer*, October 10, 1894). The Colonial Office noted that expert advice and skilled supervision were beginning to be needed for the forests of the Protected Native States (16 November 1894) and suggested that Ridley should be retained, at least for the present, and his services to be shared with those states.

Although fuming against "the science craze" and accusing Ridley of "playing the Kew game" (January 21, 1995), Governor Mitchell agreed to a temporary compromise: Ridley could stay in his present position for 1895, his salary being met partly also from Selangor state. However, he added, "I regard the office of director of forests in these small Settlements where the acreage that is—or can be—devoted to wood culture is small, as a quite unnecessary one." The state of Perak, also suggested as a contributor to Ridley's salary, refused as it was sending its own forestry officer to India for training. Governor Mitchell supported that move, remarking: "In Perak, where the preservation of forests is becoming a matter of great importance, a practical forest officer rather than a trained scientist is what is required" (Mitchell, December 27, 1894). The District Officers, annoyed at criticism from Kew, hit back with their own report on the state of the forest reserves.

In Malacca, the Assistant Superintendent of Gardens and Forests was said to have allowed a timber merchant to occupy land within a reserve and sell off the wood! In his report, the Resident Councillor observed that if a "Real Forest Department" were to be established in Malacca, the reserves would be controlled by "a specialist, chosen not for his knowledge of Botany or Natural History but a man trained in the great Indian Forest Department" (Kynnersley 1895).

Asked to spend six months in Singapore and six in Selangor, Ridley surveyed the forests of Selangor in 1895 and laid out reserves, especially in the "camphor forests." He wrote to Stephens, the Perak Forest Officer being sent for training in India and Burma: "You will see that a great deal of their method is impractible here and our system will have to be a very different one—e.g. they chiefly work with forests of one tree, we have exclusively mixed forests, which alters almost everything in the forest work plan" (Ridley 1895)

This series of exchanges, a selection from a more voluminous correspondence, illustrates the kinds of factions and personal animosities which could build up among the expatriate community of a small colony such as the Straits Settlements. Annoyance with the perceived interference of Kew in the running of the colony, intolerance with the techniques of "pure science" in a rapidly developing frontier society and inter-departmental disputes over territory are all clearly displayed. Ridley's obsession with Para rubber brought strong criticism, until he was vindicated and the crop became the mainstay of the colony. His comments on forest management were later echoed by the first generation of trained foresters in the Federated Malay States.

Control of the Straits Settlements forests was transferred to the Land Officers from 1895 to 1901, pending a more permanent arrangement. A visit in 1901 to both the Straits Settlements and the Federated Malay States from the Inspector-General of Forests to the Government of India resulted in a number of recommendations. Reserves needed to be established, especially for single species forests of mangrove and gutta percha, together with areas containing other valuable trees which were being overworked. Certain hilly lands also needed to be reserved for climatic reasons, irrespective of their forest condition. The other important recommendation was for the organization of a properly trained Forest Department (Hill 1900). An officer from the Burma service, Burn-Murdoch, was appointed Conservator of Forests to cover both the Straits Settlements (three months of the year) and the Federated Malay States. Ridley remained at the Singapore Botanical Garden, as planters were becoming interested in rubber and the Garden was the main supplier of seeds. It is ironical that Ridley, both forester and experimenter in acclimatization, should be devoting all his time to encouraging the large-scale planting of rubber. Within two decades, extensive areas of the Malayan Peninsula's lowland rain forest had been converted to the plantation crop.

From 1900 to 1920:
Burma, the American Philippines, Java and the Outer Islands, the Federated Malay States

Burma
Two main issues will be examined here: the reactions in Burma to the "Voelcker Resolution" of 1894 and activities to implement it, and the continuing question of forest access by plains people, especially those living in the Irrawaddy Delta.

The first of the forest categories suggested by the 1894 Resolution was "Protection forest to be maintained on climatic or physical grounds." This group of forests was the subject of much discussion during the first decade of the twentieth century. Large areas of new reserves were created, first in the dry zone of central Burma, then later in more remote northern districts, at least partly for ecological reasons. Such districts were inhabited by minority groups, including Shans and Kachins. As in Java, the protection of uplands invariably meant restrictions on the shifting cultivators who occupied them: "Forest officials sought to protect watersheds through reservation and new restrictions on shifting cultivators in the belief that this would safeguard plains villages from flooding and decreased water supplies" (Bryant 1997: 115). An enquiry by the government of India in 1909 produced no evidence that deforestation had in fact produced climatic or hydrological change. After much discussion the famous forester Dietrich Brandis summed it up, at least from the point of view of rainfall: "The climatic influence of well-stocked forests is limited to their immediate vicinity...*but the climate of the country cannot be changed*" (quoted in 'Op' 1911: 367). The question of water supplies and the beneficial impact of forest as preventing erosion seemed to remain more open. 'Op' suggested that "the government of Burma would do well to drop further references to the beneficial effect upon the climate of a country exercised by forests, and to confine their energies to the preservation and maintenance of forests in catchment areas and on steep slopes liable to suffer from erosion" (368). While he agreed that shifting cultivation among the Kachin hill people did harm to the forest, he believed it to be impracticable to try to abolish the practice. A kind of mixed agroforestry system was suggested, with some species of *Albizzia, Erythrina* and possibly *Alnus nepalensis,* with bamboos at lower altitudes. This more relaxed attitude to shifting cultivation could easily be maintained in a remote mountain area with little competition from other interests.

The third category in Voelcker's classification, "Minor forests to be managed in the interests of the local population," had not been present in Burma after the failure of the concept of "village forests" under the 1882 Rules. Many forests were perceived as being simply locked away from any kind of local use. "There are reserves in Burma which have been under the control of the Forest Department for thirty years in which no form of professional exploitation or organized improvement has taken place. It is neglect of this kind which has caused the undoubted dislike of the Department felt by the mass of the people." ('Op' 1911: 377). The closing of the agricultural frontier in Lower Burma, which Bryant (1997) estimates occurred about 1914, put more pressure on available forest resources. One suggestion was to bring back a variant of the *taungya* system and establish forest villages inside reserves, where people could have free access to forest resources in return for silvicultural work, aiming to increase the amount of teak or other valuable timber in the reserve. "We take up enormous areas of reserve, which contain millions of tons of unsellable material of the kind required by the population, and we prevent that population from using it" (Grieve 1916: 445).

Access to forest resources was needed even more by people on the plains and in the delta. Original plains "fuel" reserves attempted to charge for firewood but, as in Java, people simply stole to meet their needs. "The increase of population and the spread of cultivation with the concomitant destruction of accessible unprotected forests is driving the agricultural population into the reserves...stealing, often on a large scale" (*Report of Forest Administration* 1917). A scheme for making forest villages on co-operative lines in the *kanazu* reserves of the Irrawaddy Delta was recommended in 1917 by the Deputy Conservator of Forests. Each village would be provided with land for cultivation as well as a portion of the surrounding forest to look after (Lawrence 1917). After the 1914–1918 war an inspection party visited the district to consider this scheme. It was concluded that the reserves should remain for the present, as there was still timbered land available, and village settlements did not seem urgent. It was admitted, however, that the reserve system was not working. The Commissioner for the Irrawaddy district (English) made annotations to the Lawrence report. Where Lawrence deplored the burning of *kanazu* trees outside the reserves, English dismissed that as "a mere puff compared with what had gone up in smoke during the past 40 years." So many people were making a living stealing from the reserves that he concluded "reservation...was a farce," making the Government "a fool" (quoted in Watson 1920: 6). The 1920 survey also discovered huge amounts of theft "extending to over 50 per cent of the growing stock" (Watson 1920: 6). The survey concluded that a colonization scheme should remain in abeyance for the present. It was also reasoned that the policy of creating fuelwood reserves could eventually only be temporary. "Press of population will cause its revision in the interest of national economy, and the main criterion will be the relative return from agricultural as compared with forest crops" (Watson 1920: 9).

It was obvious that these delta reserves, despite the great demand for their products, had become a problem to maintain. Bryant notes that the period up to 1920 represented the pinnacle of the colonial forest department's power. Following the granting of partial self-rule to Burma in 1923, forestry passed into the control of Burmese professionals who argued that "the cause of conservation would be better served if forest management more accorded with the wishes of the people." This brought some gradual changes in attitudes (Bryant 1997: 133).

The Philippines under American Colonization

Before peace had been fully restored in the Philippines, the new Forestry Bureau was in action. Bureau Chief Ahern moved with great energy to recruit American foresters for the service and to establish forest stations in all districts. Resident foresters began to explore and map the forest resources and regulate trade in forest products. Botanist Merrill was sent to Java to investigate the techniques used there in studying flora and to examine plant collections in the herbarium and Botanical Garden at Buitenzorg (Bogor). Although the Garden had almost no Philippine material, many plants were common, so the problem of names for different genera and species began to be tackled (Merrill 1903).[26] Sherman, who explored the gutta percha trade in parts of Mindanao, also traveled to Singapore, the Federated Malay States and Java, inquiring into both the species of gutta grown and the marketing system (Sherman 1903). The isolation of the Philippines in forestry matters was quickly being reduced.

Despite much previous cutting of timber, Ahern was also sanguine that "We have an enormous area of almost virgin forest, and with careful forest management, all land more suitable for forest than agriculture should be kept under timber" (Ahern 1902: 474). Free use privileges had initially been given to the population to cut timber and forest products for their own needs, but by 1907 these were being restricted to designated "communal forest" areas (*Annual Report 1906–1907*). An early decision of Ahern was to encourage large companies to engage in logging on "exclusive licenses." It was decided that at least six companies were needed using modern methods, both to meet local demands for timber and to begin an export trade. In 1908 two large concessions "provided with modern exploitation tools" were operating: in 1914 this had risen to 11. One limitation was that to operate an exclusive license, a company needed forests with a large timber mass per ha.

American observers were interested, both in the dipterocarp timbers, which they likened to conifers because of the resin (Bryant 1907), and in the harvesting techniques and rules for working the forests, including the land question (Moore 1910). Barrington Moore had a number of comments. He regarded land as the most difficult problem, as it involved "Caingins, Cogons and Homesteads." While a law existed against making caingins or swidden plots, "so far the Bureau of Forestry has not only not been supported in its attempts to enforce the law, but has *actually been prevented* from doing so." (Moore 1910: 80). Cogons, or *Imperata* grasslands, would remain untouched because claims "by caciques" usually existed over the land, while homesteads "have been for the most part, nothing but mere Caingins." Establishing reserves to concentrate the work had been adopted as "the keynote of the whole policy of the Bureau" (p. 154), but the vegetation and land use of the entire islands would first need to be mapped, a huge task (p. 151). In summing up the influence of the forests, Moore began by invoking the usual hydrologic and desiccationist myths: "There is ...strong evidence to show that forests not only regulate the run-off and retain water in the soil, but actually influence the total quantity of rainfall as well." His further comment is more apposite: "In the Philippine Islands, as in no other country, does the solution of the forest problem involve the solution of the land question. Upon the proper handling of this question depends the agricultural development of the country and hence the welfare of a people almost wholly dependent on agriculture" (p. 154).

The activities of the Bureau of Forestry attracted considerable attention from neighboring countries. H.J. Kerbert, a forester from the teak district of Jepara in Central Java, visited the Philippines and discussed his findings in the new journal of the Netherlands East Indies Forest Service, *Tectona*. In his first paper Kerbert stressed the importance of the dipterocarp forests, on which the development of forest industries to a large extent depended (Kerbert 1908: 527). He also mentioned the activities of the Insular Lumber Co., operating on Negros with highly mechanized techniques. In his second *Tectona* paper, Kerbert used the Philippine example to stir up the Java-focused establishment to do more about regulating the timber industry of the Outer Provinces of the Netherlands East Indies. He warned that the large-scale production from 11 big companies operating in the Philippines would soon offer competition to timber from areas such as the *panglong*[27] in Riau and suggested that a number of aspects of the Philippine system could be adopted (Kerbert 1914).

Also interested was Pearson, from the embryonic Forest Department of British North Borneo, who visited the lumber camp of Cadwallader and Gibson in steep country at Bataan, 30 miles from Manila. He was impressed as he watched logs being clear felled on the slopes and shifted by winch and donkey engine, sawn and transported to a wharf at tidewater, in an entirely mechanized operation. Reports of the damage caused by these kinds of operations surfaced, however, from time to time in the *Annual Reports* of the Bureau of Forestry. In 1914 the difficulties of reproducing desirable species on the areas logged over by steam techniques were discussed. As the forest crown was destroyed, the intense insolation was too severe for most seedlings. Ripping out the larger trees also damaged the rest of the stand.

While noting the competition likely to be offered to North Borneo by the Philippine product, Pearson made some interesting comparative

Photo 2-4. Developing a lumber concession in the Philippines. Source: "Annual Report of the Director of Forestry of the Philippine Islands for the fiscal year ended June 30, 1911"

comments about forestry in the Philippines and the Federated Malay States. "Whereas in the Federated Malay States, the practical administration is effective and the scientific investigation sketchy, the reverse may be said of the Bureau here" (Pearson 1914: 4). He went further "The general policy of the Bureau differs entirely from that of the Malay States. In the latter country many large reserves are created, which are either closed altogether and "improved" or else opened for exploitation under restrictions as to minimum girth, species to be cut etc. In the Philippines the Director of Forestry holds an extraordinarily strong position...In the Malay States all public land is at the disposal of the Lands Department until surrendered to the Forest Department while in the Philippines all land is at the disposal of the Forest Department until it is surrendered to the Lands Department" (Pearson 1914: 9). The problem in the Philippines was the slowness of the approval for licenses, both for timber cutting and agricultural land, because of a lack of trained staff to perform the necessary checks.

The land question was a complicated one, especially that relating to homesteading. A "homestead" (16 ha) presumed a permanent plot, presumably for dry farming. Such plots, with no capital to work them, were not usable by Filipino households. Most "public" land was located in remote regions, and some was probably swidden fallow, not recognized as having continuing ownership. Some of those receiving "homesteads" were foreigners, such as Japanese abaca growers in Davao (Mindanao). They acquired lands belonging to the indigenous Bagobo, who then attacked them (Hayase 1984). When a new public land law, passed in 1919, made homesteading easier, there were contradictory reports of its success. In some districts, *cogon* lands were being rapidly settled as homesteads, and just as rapidly being abandoned. People who could not control the grass were invading nearby forests, leaving their former homesteads to be used for grazing and opening new *caingin* for their subsistence (*Annual Report* 1920). In 1920 an area of 182,892 ha. had been examined by officers of the Bureau of Forestry for their agricultural potential, together with 3172 individual public land applications, which had been inspected and certified to the Bureau (Nano 1951). The homestead idea, very much an American concept, was not very appropriate to the conditions prevailing in the Philippines, especially in the so-called "Non-Christian areas," where exploitation of indigenous minorities was common.

A process of "Filipinization" was in place in the Forest Service, beginning with the founding (in 1910) of the Philippine Forest School at Los Banos. In 1914, Filipinos were reported to be in charge of nine of the 11 forest districts into which the islands were divided. Nevertheless, the

Bureau of Forestry was not popular. Both forest reserves (of which a few were set aside) and the release of large tracts for steam logging to outside (mainly American) companies were opposed by local interests. It was the familiar story of colonizers wanting to change local cultures, carrying with them the crusading message of forest protection. Dean Worcester's attitude is typical.

Thus far the Filipinos have made no attempt to share in the development of their forests on any save a very small scale...So far as concerns conservation, the attitude of the Filipinos is even less satisfactory. There is abundant evidence on which to base a prediction as to the policy they would follow in practice, if the compelling hand of an enlightened nation were withdrawn (Worcester 1930 [rev.]: 605).

Java and the Outer Islands

Debates continued in Java over the impact of forests on both climate and hydrology, but it was not until the late 1920s that serious attention began to be given to the uplands. Large scale reforestation began in that decade, its pace quickening after the Forest Law of 1927 which limited forest conversion. In 1937 all remaining non-teak forest on Java was legally protected.

In the early years of the twentieth century, more serious thought began to be given to the Outer Islands forests, which had been in charge of the

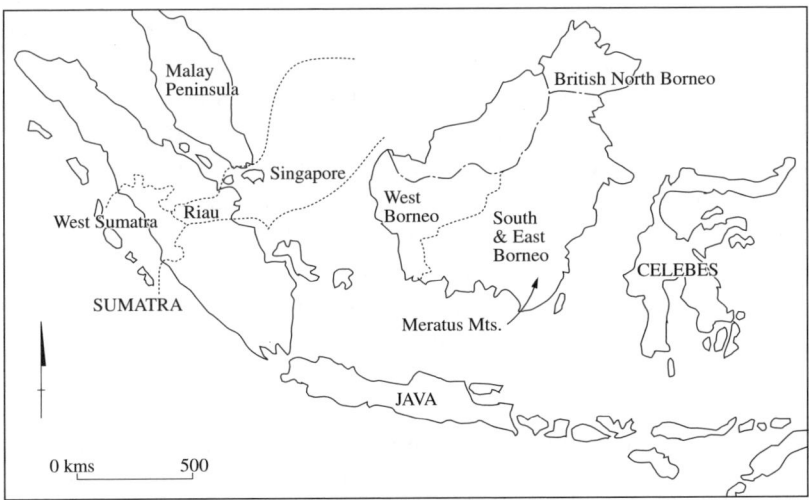

Figure 2-5. Java and the outer islands

general administrative officers in the districts concerned. In 1910,Tobi introduced the possibility of large-scale steam logging and sawmills, along Philippine lines, being established in Sumatra (Tobi 1910). The question was examined quite thoroughly, following Kerbert's first paper, but rejected on the grounds of unsuitability, especially in the swampy *panglong* area (Lovink 1910; Treub 1910). In the same year an investigation was undertaken in West Sumatra on the impact of mountain deforestation on agricultural conditions in the lowlands (Herwerden 1916). The resulting report led to a skeleton forest warden staff being appointed in Sumatra and Celebes in 1912, while the famous "Jelutong question"[28] resulted in a similar development in South and East Borneo in 1915. Despite these preliminary efforts, forest services beyond Java were just starting to become organized by 1920. Forest produce trade continued to be focused on non-wood materials rather than timber. Smallholder rubber was making inroads, however, resulting in extension of permanent cultivation in some districts of Kalimantan and Sumatra. It was not until the 1930s that there was much interference with local forest activities, through regulation of the timber trade or setting up of forest reserves (Potter 1988).

The Federated Malay States

To a forester from Burma, the initial impression of the evergreen forests of the FMS was one of the enormous size of some of the trees and the large number of species, comparatively few of which were valuable (Burn-Murdoch 1904). Such sentiments were echoed a decade later by Barnard, then Acting Conservator of Forests following Burn-Murdoch's death. According to Barnard "the very great differences which exist between the forests of India and those of the Federated Malay States would give an Indian forester no advantage in this country." While in India most of the forests were deciduous, with comparatively few species, "the Federated Malay States forests on the other hand are dense and evergreen, and the number of the species of trees is very large. It is evident that the same system cannot be applied to both" (Barnard 1914). The girdling and drying methods applied to teak, for example, would not do in the rain forest. As Ridley had noted in the 1890s, new techniques would be necessary to handle these kinds of forest.

However, Burn-Murdoch began in the traditional manner, by laying out reserves. These were initially for gutta percha and mangroves, following Hill's recommendation, but they were also the easiest to deal with, as the species were concentrated in specific areas. He also gave serious thought to the management of the dipterocarps, selecting those with the best hardwood timber, such as *merbau* (*Afzelia palembanica*) and *chengal* (*Balanocarpus maximus*) and making reserves in places where those trees were well represented. In 1904 the largely Chinese-owned tin mines were the biggest consumers of wood. They viewed the resource as inexhaustible, resisting efforts at taxing or controlling logging (Burn-Murdoch 1904). There was no mention of other occupants of these forests, such as the various groups of Orang Asli[29], who were active collectors of forest products and claimants to scattered fruit holdings, especially of durians (Harper 1997). The rapid increase in rubber prices quickly produced another, stronger competitor for forest land.

Two booms occurred in rubber prices, in 1905–1906 and 1908–1910, as the automobile industry became established in the USA and demand for rubber tires rose quickly. The area under rubber in the FMS rose from less than 8,000 ha in 1904 to over 160,000 in 1912 and almost 500,000 in 1922. While early plantings had largely been confined to the central west coast (Perak, Selangor and Negri Sembilan), from 1907 they increased in the south also (Malacca and Johore), so that by 1922 the total area alienated was 932,000 ha. (Drabble 1973: Appendix III). The eastern districts, including Pahang, were much less involved, largely because transportation tended to be concentrated on the western side of the peninsula. In the early period, rubber was interplanted with other crops, but increasingly, plantations were laid out in forest areas, especially on the slopes of the main range.[30] A report in the *Straits Times* provides a graphic illustration.

> Traveling through Malaya during the past year one came across scores of the black smoke columns which mark the chief stage of jungle clearing. But far more weirdly impressive is the aspect of the new plantations. Vast areas are seen on which there is no scrap of vegetation. Gaunt skeletons of the few trees which have been left to serve as landmarks stand out against the sky. The black earth is thickly strewn with the whitening trunks of felled trees, and they look like the rainwashed bones of great giants which have fallen in a dreadful battle. The old splendor of the jungle has passed away as if a mighty cataclysm of nature had destroyed it, and the softer, more regular beauty of the plantation has yet to come. (*Straits Times* August 24, 1912)

The faster the rubber industry expanded, the more feverishly the foresters worked to create reserves. In 1904 only one per cent of forests were reserved: this had risen to 11 per cent in 1908 and 32 per cent in 1913 (Allured and Sallyanne 1934). Despite these efforts, the loss of timber in clearing for rubber plantations was enormous, estimated by Foxworthy (1921) at 50 million tons during the previous 20 years. In a paper written in 1930, Barnard looked back on the scene: "There was, of course, some hostility to reservation from those who entered the struggle for land too late ... had not the reserves already been in existence, it would have been difficult for the government to have resisted the incessant clamor for land. Even as it was, the destruction of valuable timber was prodigious" (Barnard 1930).

Despite the creation of Malay Reservations in 1913 (partly to prevent Malays selling their land to entrepreneurs) and efforts to confine the Malay population to rice production, one third of the area under rubber in 1921 was controlled by smallholders, a large percentage of whom were Malay (Nonini 1992). Some of this cultivation was undertaken by squatters on state land and at least a portion was found by the 1930s to be grown by Orang Asli inside the forest reserves, where they had cultivation privileges (Harper 1997). This last discovery was embarrassing to the authorities. They had considered the aboriginal population as incapable of involving themselves in commercial activities, yet had encouraged them to adopt settled cultivation.

Most of the more common dipterocarp species were not considered valuable, being used locally largely for firewood, but also in mining, railways and rubber factories. Unlike the position in the Philippines, there was no attempt to develop an export trade in timber, as it was anticipated that production for the entire Malayan Peninsula would barely meet requirements (Foxworthy 1921). Most of the reserves were not exploited, many fulfilling also a protective function on steep mountains (Cubitt 1921; Barnard 1930). However, the American influence was there[31] and considerable controversy arose as to the suitability of such mechanized and intensive methods for the Malayan situation. Eventually the idea was rejected, the European "conservatively organized sources of permanent... raw material" being compared favorably to the American "fugitive and destructive" type (Oliphant 1934: 46). Even British North Borneo eventually gave up the American model on economic grounds, after trying it for a few years.

Conclusion

In this paper I have sought to outline a number of the political, ecological and economic factors which influenced the activities of colonial forest services (or their precursors), in Southeast Asia between 1875 and 1921. I have argued that the Franco-German "scientific forestry" model was initially accepted in teak areas of Java and Lower Burma; in Burma the system was the focus of experimentation on the forest reserves. The attempted exclusion of human populations and the myopic specialization on individual trees made its actual operation more difficult, and subjected to much compromise with local groups and their supporters. In both of these cases, teak regeneration occurred with the help of local people, though this broke down in the case of Burma when the enormous reserves of Upper Burma became available. Outside the teak producing localities, which were commercial in orientation[32] and the main concern of the respective forestry institutions, other forests received different treatment. They were considered expendable where faced with competition from commercial crops, but were at times also regarded as worth preserving for climatic or hydrologic reasons. In the case of land use competition, plantations were inevitably preferred over indigenous swidden systems. Swidden cultivators were castigated largely in ecological terms, in the Philippines as well as Java, as being the direct cause of whatever environmental problems might be observed.

Not only were the foresters strongly and universally disliked, despite their lofty statements regarding the "value" of the forests, but the people themselves made their own accommodations, both stealing timber and sometimes, as in the Burmese case, treating the delta reserves as "a joke." One may argue that they were forced to adopt such behavior because of the ways in which they were deprived of land and resources by colonial authorities. The stereotypes held by foresters and others about the motivations and abilities of certain groups, together with the remoteness of many of the forests, also provided a cover for "aberrant" activities to go undetected, such as Orang Asli rubber planting in reserve forests.

Among the British colonial territories, the reserve system actually seemed to work best in Malaya, where it was introduced considerably later than India or Burma. In Malaya the percentage of total land area occupied by forest reserves by the late 1930s was 19.7, compared with 13.7 in Burma and 10.3 in India. Comparable figures for North Borneo were 1.6 per cent and Sarawak, four per cent (Troup 1940). It is clear that the forestry department in Malaya was both stubborn and strong in its insistence on this "locking up" of valuable land as forest reserve. Despite the statements of foresters and others that the system practiced in Burma could not work in the diverse tropical rain forest, the same reserve system was established, though silvicultural techniques were different. The links with India and Burma among forest department personnel also predisposed them to look for similarities with the "tried" system, despite differences in conditions.

The lack of an export focus to timber harvesting also allowed time to develop systems more suited to the nature of the dipterocarp forests.

The Spanish foresters in the Philippines, being largely cut off from ideas on forest management and short of personnel, were unable to evolve a system which combined the commercial needs of the market with their conservation ethic. This left a silvicultural "tabula rasa" on which the incoming Americans could experiment. Coming from a more capitalistic tradition of privatized forests, they were able to focus on a few large firms, developing timber exports at a level of mechanization which amazed their neighbors. The Dutch, in their almost total fascination with Java and teak, found it difficult to do much about the Outer Islands forests, but they did consider and reject the American system. The existence of the exciting alternative, rubber, meant that Malaysia could afford to be conservative in its approach to the forests: the mechanized model was rejected there as well.

While environmental ideas about the influence of forests on climate and hydrology continued to be invoked, such ideas were subjected to more criticism later in the period and were gradually losing credence. Even in the 1870s, however, such ideas seem to have been more useful in the moral pressure they could exert, than productive of any ecological benefit. They would usually give way to economic determinants in the end.

Despite the similarities which have been uncovered, it is obvious that each colony was different in its approaches to its forests, and that these approaches also changed over time. As shown in the Malayan case, the idea of a forestry department exercising control over commercial interests in the name of "science" was anathema to local administrators in the 1880s. Yet by 1910, the forest administration was perceived as strong in practical terms and weak in science. While in Burma and Malaysia it was considered inevitable that most forests would be replaced by agriculture, this was not the thinking in the Philippines. The Philippines managed to export dipterocarp timber, but their system of allocating land proved inadequate and inequitable. By 1920, land scarcity was being felt in the Burma delta and in upland Java, where forests were placed under extreme pressure. In contrast and despite the plans of colonial authorities, smallholder rubber was becoming a success in both Malaya and the outer islands of Indonesia.

Acknowledgments

I am indebted to Professor Anthony Reid and the Economic History of Southeast Asia Project at the Australian National University (ANU), which first set me to explore the forest history of Southeast Asia and granted me a year's secondment at ANU. The University of Adelaide provided me with study leave, small grants and teaching release to pursue these interests in breaks from the normal hurly-burly of teaching and administration. I have been privileged to work on Burma records at the India Office library, the libraries of LSE and SOAS in London, the Oxford Forestry School and the National Library, Canberra; on Java and other parts of Indonesia in the Dutch National Archives at the Hague, the KITLV library, Leiden, Cornell University library, the National Archives, Jakarta and the ANU library; on the Philippines at the Escuela de Montes, Madrid, the National Library of the Philippines, Manila, the forestry library, University of the Philippines at Los Banos, the library of the University of California at Berkeley, the Yale University library and the National Library, Canberra; on Malaysia in the Public Record Office, Kew, the Kew Gardens Archives, the Singapore-Malaysia collection of the National University of Singapore, the Forest Research Institute of Malaysia (FRIM), and the National Library, Canberra.

Notes

1 These species are both highly resistant to borers and able to stand submersion in salt water.
2 *Taungya*, in the older Karen meaning, refers to "farm." It also refers to an agroforestry system in which food crops are grown as a nursery for tree crops. Both meanings are used interchangeably in this chapter.
3 One circular to forest officers concerning Karen agriculture advised: " These people are almost totally occupied with their own cultivation which requires much more time and labor than the cultivation on the plains...they have but few wants and less money and do not therefore benefit the general trade to any appreciable extent. They only become of importance when they come into contact with the Forest De-partment or the timber traders." (Conservator of Forests, British Burma to all divisional officers, 29.1.1877) (RFA)
4 Grant (1932) noted that the most rapid expansion of area under rice cultivation occurred after 1870.
5 See Grove (1995: Chapter 8).
6 "Shall we follow then the example of those improvident populations, who, by clearing of forests, diminished most unduly the annual fall of rain, or prevented its retention?" (F. von Mueller, Melbourne, "Forest culture in relation to industrial pursuits"; quoted in Kurz [1875: 27]).

7 He suggested (p. 74) that the Karen population of Burma cut each year an amount of teak timber equal to the entire English fleet! He nevertheless described their diverse swidden cultivation in some detail, and was appreciative of their forest knowledge.
8 In Adas' description of the deltaic forests, he describes tall *kanazo* forests as being widespread (1983: 98), while Kurz refers to them mainly as littoral, in strips along the creeks. A statement in the Watson report (1920) indicates that *kanazo* forests did occur along the creeks, but in the far southern part of the delta they coalesced to form extensive forests. Those areas were still forested in 1920, having been reserved between 1895 and 1901.
9 The service had previously been abolished in 1826
10 This system is called in Dutch *boschakkerbouw* or *bosveldbouw* ("forest field farming").
11 A translation of Buurman's pamphlet describing the system was published in the *Indian Forester,* but not until 1892 (*Indian Forester* vol. XVIII no 8, titled "Teak cultivation in Java")
12 Under the *Domeinverklaring*, the Agrarian Law of 1870, all land not under permanent cultivation or under fallow for more than three years was declared the property of the state (Peluso 1992: 64).
13 By comparison, in 1865 the Forest Service in Java had an establishment of two inspectors, 13 conservators, 23 rangers, 44 foresters and 110 forest guards (Boomgaard 1994)
14 The VOC, or Dutch East India Company, collapsed at the end of the eighteenth century.
15 Quoted in Ahern (1901)
16 In 1860 the port of Cebu was opened to world trade, leading to greatly increased economic activity on the island.
17 These reserve documents, stored in the India Office Library, London, contain many interesting details of population and land use in the areas concerned, together with maps and transcripts of interviews with villagers.
18 Of these, only 849 square miles were teak taungyas
19 A more complete picture of Southeast Asian forests would include Thailand and the three countries which comprised French Indochina: Vietnam, Laos and Cambodia. Restrictions of space make this impossible.
20 Singapore, Penang, Malacca, Province Wellesley.
21 Perak, Selangor, Pahang, Negri Sembilan: before 1896 these were called the "Protected Native States."

22 Another fine example of desiccationist writing!
23 Most of Singapore's timber at the time came from the Chinese-run *Panglong*, woodcutters' camps usually located along the coast of Riau.
24 See Brockway (1979). There was no forestry training school in Britain at the time; training was done in Europe or in India.
25 The discussion in the succeeding paragraphs is based on correspondence in unpublished Colonial Office files, CO273—'Straits Settlements Original Correspondence, 1838-1900,' 1893 to 1895
26 Ahern believed that native tree species numbered 600-700.
27 See footnote 17 and Colombijn (1997).
28 Jelutong was a wild rubber which experienced a boom from 1910 to 1912 in various parts of Borneo. Along the southern coast there was concern about over-tapping and tree destruction, which induced the local authorities to hand over large areas to an American concession, then employ a special police force to protect the boundaries and keep out tappers, some of whom were jailed. Due to the involvement of a local trading company and much lobbying, it became a "cause celebre" and was debated in the Dutch parliament (see discussion in Potter 1988).
29 Orang Asli, meaning "original people", has since the early 1950s been the official term for the 20 or so indigenous ethnic minority groups in the Peninsula.
30 Harper notes that the "central dynamism"of the colonial economy came from the forest frontier, with plantation interests invading precisely those areas previously settled by Orang Asli (Harper 1997).
31 Foxworthy, an American, had been appointed as Forest Research Officer to the FMS
32 They were export-oriented in Burma, less so in Java, as local needs absorbed much timber in this heavily populated island.

Bibliography

Adas, M. 1983. Colonization, commercial agriculture and the destruction of the deltaic rainforests of British Burma in the late nineteenth century. In R.P.Tucker and J.F. Richards, eds., *Global Deforestation and the Nineteenth Century World Economy*, 95–110. Durham: North Carolina University Press.

Ahern, G. 1901. Compilation of notes on the most important timber tree species of the Philippine islands. Philippine Forestry Bureau, Microfilm, Roll #264a, Philippines National Library, Manila.

———— 1902. *Report of the Bureau of Forestry of the Philippine Islands from July 1, 1901 to September 1, 1902.* Report of the Philippine Commission, Bureau of Insular Affairs, Washington.

Allured and Sallyanne. 1934. Les forets et le service forestier de la Malaisie Britannique. *Les Services Forestiers a Java et en Malaisie.* Report to French Indochina Forest Service.

Annual Report of the Director of Forestry of the Philippine Islands. 1906–1907; 1914; 1920.

Annual Report of the Straits Settlements, 1886.

Barnard, B. 1914. Forest Administration, General. B. Barnard, Acting Conservator of Forests to Under-Secretary, Federated Malay States, 8.4.14, CO273/410, No 22557.

———— 1930. Forestry in the Malay Peninsula. *Indian Forester* (April): 191–196.

Becking, J.H. 1928. *De Djaticultuur op Java: een vergelijkend onderzoek naar de uitkommsten van verschillende verjongingsmethoden van de djati op Java.* Wageningen: Veenman.

Blanco, Fr. 1837. *Flora de Filipinas* (quoted in Ahern, 1901).

Boomgaard, P. 1994. Colonial forest policy in Java in transition 1865–1914. In Robert Cribb, ed., *The Late Colonial State in Indonesia: Political and Economic Foundations of the Netherlands Indies 1880–1942*, 117–138. Verhandelingen van het Koninklijk Instituut voor Taal-Land en Volkenkunde 163. Leiden: KITLV Press.

Brockway, L.H. 1979. *Science and Colonial Expansion: The Role of the British Royal Botanic Gardens.* New York: Academic Press.

Bryant, R.C. 1907. Some Philippine Forest Problems. *Proceedings of the Society of American Foresters* II(3): 3–19.

Bryant, R.L. 1992. Political Ecology, an Emerging Research Agenda in Third World Studies. *Political Geography* 11(1): 12–36.

———— 1997. *The Political Ecology of Forestry in Burma 1824–1994.* Honolulu: University of Hawaii Press and Delhi: Oxford University Press

Burn-Murdoch, A. 1904. Notes from the Federated Malay States. *Indian Forester* (October): 458–463.

Buurman, W. 1892. Teak Cultivation in Java. *Indian Forester* XVIII: 8

Cady, J. 1958. *A History of Modern Burma.* Ithaca: Cornell University Press.

Cantley, N. 1883. Report on Forests in the Straits Settlements. Sections reproduced in the *Indian Forester* XI(10): 488–491, 1885.

Ciancio, O. and S. Nocentini. 2000. Forest Management from Positivism to the Culture of Complexity. In M. Agnoletti and S. Anderson, eds.,

Methods and Approaches in Forest History, 47–58. IUFRO 3 Research Series. Wallingford: CABI Publishing and International Union of Forestry Research Organizations.

Colombijn, F. 1997. The Ecological Sustainability of Frontier Societies in Eastern Sumatra. In P. Boomgaard, F. Colombijn, and D. Henley, eds., *Paper Landscapes: Explorations in the Environmental History of Indonesia*, 309–340. Verhandelingen van het Taal-land-en Volkenkunde 178. Leiden, KITLV Press.

Cordes, J.W.H. 1881. *De Djati-Bosschen op Java; hunne natuur, verspreiding, geschiedenis en exploitatie.* Batavia: Ogilvie & Co.

Cubitt, G.E.S. 1921. Proposed Malayan Forest Service. Encl No 4 in 352/21, CO 874/717.

Drabble, J.H. 1973. *Rubber in Malaya: 1876–1922: The Genesis of the Industry.* Kuala Lumpur: Oxford University Press.

Fenner, B.L. 1985. *Cebu under the Spanish Flag, 1521–1896: An Economic-social History.* Cebu City: San Carlos Publications.

Foxworthy, F.W. 1921. *Malayan Science Bulletin 1—Commercial Woods of the Malay Peninsula.* Kuala Lumpur: FMS Forest Department.

Gooszen, H. 1999. *A Demographic History of the Indonesian Archipelago, 1880–1942.* Leiden: KITLV Press.

Gorkum, K.W. van. 1874. *Een woord over de ontwouding van Java ter ontginning van gronden.* (TNL in Nederlandsch Indie, 2, 1874, Deel XIX, Aflevering III 382–421).

Grant, J.W. 1932. The rice crop in Burma. *Agricultural Survey* 17. Rangoon: Government Printing and Stationery.

Grieve, J.W.A. 1916. Note on Forest Policy in Burma. *Indian Forester* XLII(9): 444–447.

Grove, R. 1995. *Green Imperialism: Colonial Expansion, Tropical Island Edens and the Origins of Environmentalism, 1600–1860.* New Delhi: Cambridge University Press.

Haeuber, R. 1993. Indian Forest Policy in Two Eras: Continuity or Change? *Environmental History Review* 17(1): 49–75.

Ham, S.P. 1895. De Instandhouding en uitbreiding der Climaatbosschen en de Verhooging van de productiviteit van de Indischen bodem. *Tijdschrift voor Nijverheid en Landbouw in Nederlandsch Indie.* Deel L, Aflevering VI: 351–385

Harper, T.N. 1997. The Politics of the Forest in Colonial Malaya. *Modern Asian Studies* 31(1): 1–29.

Hayase, S. 1984. Tribes, Settlers, Administrators on a Frontier: Economic Development and Social Change in Davao, Southeastern Mindanao, the Philippines, 1899–1941. Paper presented to the Asian Studies Association of Australia Fifth Annual Conference, Adelaide, May 13–18.

Hefner, R. 1990. *The Political Economy of Mountain Java*. Berkeley: University of California Press.

Herwerden, G.A.M.H. van. 1916. Eenige beschouwingen over de bestrijding van den roofbouw der Inlandsche bevolking in Nederlandsch Indie. *Koloniaal Tijdschrift* 5(1): 9–16,145–164.

Hill, H.C. 1900. Report on the present system of forest administration and management in the Federated Malay States and suggestions for the future management of the forests of those states. Her Majesty's Indian Forest Service, Printed at Selangor Government Printing Office.

Holle, K.F. 1892. Maatregelen tegen verspilling van den bouwgrond, tegen roofbouw en uitputting van den bodem. *Tijdschrift voor het Binnenlandsch Bestuur*, deel 7: 74–90.

——— 1894. Eenige gegevens omtrent den inlandschen landbouw op tegalens en het nut van terrassen. *Tijdschrift voor Nijverheid en Landbouw in Nederlandsch-Indie* XLVII, Aflevering 1: 1–13.

Jewitt, S. 1995. Europe's "Others"? Forestry Policy and Practices in Colonial and Postcolonial India. *Environment and Planning D: Society and Space* 13: 67–90.

Jordana y Morena, R. 1876. *Memoria sobre la produccion de los Montes Publicos de Filipinas durantes el ano economico de 1873–4*.

——— 1891. *Estudio Forestal acerca de la India Inglesa, Java y Filipinas*. Madrid: Moreno y Rojas.

Junghuhn, F. 1850–1854. *Java, deszelfs gedaante, bekleeding en inwendige structuur*. Amsterdam: van Kampen.

Kerbert, H.J. 1908. Amerikaansch Bosbeheer op de Philippijnen. *Tectona* I & II: 377–394, 517–546.

——— 1914. De Ontwikkeling van de houtindustrie op de Philippijnen in verband met die in onze Buitenbezittingen. *Tectona* 7: 607–615.

Koorders, S.H. 1895. Spontane en Kunstmatige Reboisatie van den Sendoro op Java. *Tijdschrift voor Nijverheid en Landbouw in Nederlandsch-Indie* XLIX, IV, V & VI: 241–287.

Kurz, S. 1875. *Preliminary Report on the Forest and other Vegetation of Pegu*. Calcutta: Baptist Mission Press

Kynnersley, C.W.S. 1895. Minute by the Resident Councillor, Malacca 22 March 1895. Enclosure in Despatch No 240, Governor Mitchell to the Marquess of Ripon, 14 June 1895, CO273/204.

Lawrence, A. 1917. *Report on the Suitability of Certain Portions of the*

Tidal Reserves in the Myaungma and Pyapon Districts for Cultivation (Burma collection, Oxford Forestry School library).

Lovink, W.G. 1910. Letter in 'Verbaal' 15.07.1910, No 15, and 'Indische Depeche' 17.5.10, No 542/3. ARA Den Haag.

Lowood, H.L. 1990. The Calculating Forester: Quantification, Cameral Science and the Emergence of Scientific Forestry Management in Germany. In T. Frangsmyr, J.L. Heilbron and R.E. Rider, eds., *The Quantifying Spirit of the Eighteenth Century,* 315–342. Berkeley: University of California Press.

Lugt, C.S. 1933. *Het Boschbeheer in Nederlandsch Indie.* Haarlem: Willink.

Merrill, E.D. 1903. Report on Investigations Made in Java in the year 1902. *Forestry Bureau—Bulletin* 1. Manila: Bureau of Public Printing.

Moore, B. 1910. Forest Problems in the Philippines. *American Forestry* XVI(2–3): 75–81, 149–154.

Nano, J.F. 1951. Brief History of Forestry in the Philippines. *Philippine Journal of Forestry* 8: 1–4 *and* 9: 113.

Nisbet, J. 1901a. *Burma under British Rule and Before* Vol.I. London: Constable and Co.

——— 1901b. *Burma under British Rule and Before* vol. II. London: Constable and Co.

Nonini, D.N. 1992. *British Colonial Rule and the Resistance of the Malay Peasantry, 1900–1957.* Monograph Series 38. New Haven: Yale University Southeast Asia Studies.

Oliphant, J.N. 1934. Some Economic Aspects of Timber Production in Malaya. *Empire Forestry* 13(1): 45–57.

'Op'. 1911. Want of a Definite Forest Policy in Burma. *Indian Forester* XXXVII (July): 364–380.

Owen, N. 1971. The Rice Industry of Mainland Southeast Asia 1850–1914. *Journal of the Siam Society* 59(2): 75–143.

Palte, J.G.L. 1989. *Nederlandsche Geografische Studies No 97 Upland farming on Java, Indonesia: A Socio-economic Study of Upland Agriculture and Subsistence under Population Pressure.* KNAG & Utrecht: Geografisch Instituut Rijksuniversiteit Utrecht.

Pearson, H.C. 1914. Memorandum on forestry in the Philippines CO874/711 (Annual Reports, British North Borneo).

Peluso, N.L. 1992. *Rich Forests, Poor People: Resource Control and Resistance in Java.* Berkeley: University of California Press.

Peluso, N.L. and P. Vandergeest. (in press). Genealogies of the Political Forest and Customary Rights in Indonesia, Malaysia and Thailand. *Journal of Asian Studies.*

Potter, L. 1988. Indigenes and Colonizers: Dutch Forest Policy in South and East Borneo (Kalimantan), 1900 to 1950. In J. Dargavel, K. Dixon and N. Semple, eds., *Changing Tropical Forests: Historical Per-spectives on Today's Challenges in Asia, Australasia and Oceania*, 127–152. Canberra: Centre for Resource and Environmental Studies, ANU.

———— 1997. A Forest Product out of Control: Gutta Percha in Indonesia and the Wider Malay world, 1845–1915. In P. Boomgaard, F. Colombijn and D. Henley, eds., *Paper Landscapes: Explorations in the Environmental History of Indonesia*, 281–308. Verhandelingen van het KITLV 178. Leiden: KITLV Press.

Report on the Forest Administration in Burma for the year ended 30 June 1917.

Report on the Forest Administration of British Burma, 1877.

Resolution on the Forest Administration of British Burma, 1881–1882.

Revista de Montes, 1877–1898, Madrid: Colegio de Montes.

Ridley, H.N. 1895. Ridley to Stephens, quoted in 'Federated Malay States. Forest Conservancy' 40314, No 460, CO 273/263, 1900.

Roth, D. 1983. Philippine Forests and Forestry: 1565–1920. In R. Tucker and J. Richards, eds., *Global Deforestation and the Nineteenth Century World Economy*. Durham, N.C.: Duke Press Policy Studies.

Sherman, P.L. 1903. *The Gutta Percha and Rubber of the Philippine Islands*. Department of the Interior, Bureau of Government Laboratories. Manila: Bureau of Public Printing.

Singapore Free Press, 7.11.1893 (enclosed with correspondence in CO273).

Sivaramakrishnan, K. 1996. The politics of fire and forest regeneration in colonial Bengal. *Environment and History* 2: 145–194.

Stott, P. and S. Sullivan, eds. 2000. *Political Ecology: Science, Myth and Power*. London: Arnold.

Straits Times, 1900–1920, various issues.

Tobi, E. 1894. Reboiseering van den Goenoeng Sendoro en Goenoeng Soembing. *Tijdschrift voor Nijverheid en Landbouw in Nederlandsch-Indie* XLVIII, aflevering III: 147–151.

———— 1910. Government Forests of Java. *Indian Forester* XXXVI(1–2): 106–108.

Treub, W.G. 1910. Letter in 'Verbaal' 15.07.1910, No 15, 'Mailrapport' 1909, 1802. ARA, Den Haag.

Troup, R. 1940. *Colonial Forest Administration*. Oxford: Oxford University Press.

Vidal y Soler, D. 1877. *Manual del maderero en Filipinas.*, Madrid: Imp. de la Revista Mercantil, Manila.

Vidal y Soler, S. 1874. *Memoria sobre el Ramo de Montes en las Islas Filipinas.* Madrid: Aribau y Cia.

Watson, W. 1920. *Notes on Inspection of Forest Reserves in the Pyapon and Myaungmya Districts of the Irrawaddy Delta.* Burma collection, Oxford Forestry School library.

Worcester, D.C. 1930. *The Philippines, Past and Present.* New edition by Ralston Hayden. New York: MacMillan.

3
Trading in the Forest: Lessons from Lao History

Deanna Donovan

With a relatively large area of natural forest still intact, Lao PDR[1] is unique among the countries of Southeast Asia. The forest contributes significantly to the national economy in addition to playing an important role in providing basic necessities for the largely rural populace. Forestry accounts for 10.65 per cent of agricultural contribution to Gross Domestic Product. As a source of government income, timber royalties rank third providing 12 per cent of total tax revenues in 1998/99 (Ministry of Finance 2000). Forest product exports consist not only of lumber and plywood, but also many non-timber forest products, including cardamom, rattan and benzoin, for which Laos is historically famous. In terms of income-earning potential, these products hold great promise if properly managed. However, open access to most forest areas and increasing demands from a rapidly growing domestic market as well as foreign exporters have reduced forest cover by almost half over the past century, from nearly 70 per cent in 1941 to the approximately 41 per cent today (Foppes and Ketphanh 2000). There is little obvious evidence though that the proceeds of the conversion of forest assets have been invested in the nation's socio-economic development. Few measures appear to have been taken to insure a comparable flow of forest resources well into the future. Thus, unfortunately it appears that the country may be liquidating its natural capital for the mere facade of modernity with very little substantial advancement for most of the population.

The purpose of this paper is to assess the role forest products have played in the national development of Laos. It will discuss how the trade in forest products appears to have benefited some groups as opposed to others, and how this is reflected in both socio-economic development and environmental conditions as a whole. It shows how patterns of forest exploitation endure through time with all too predictable results. The

Photo 3-1. A *styrax* tree tapped for benzoin resin in Laos. Photo by Satoshi Yokoyama.

paper begins with a brief discussion of the analytical framework adopted and a review of research methods. Following an overview of the physical setting and current socio-economic conditions in Laos, it examines the country's economic and political evolution and explores the role of trade as a key factor in development, both nationally and regionally. Next it examines the pattern of trade over time and assesses the influence of trade on environmental and socio-economic conditions in Laos. Finally it looks at the implications of current trends and suggests some lessons to be learned from the Lao experience of trading forest products.

Research Methods

In this analysis of Lao history focus is on the exploitation and trade of forest products with the objective of highlighting the role these resources have played in national development. Following a mode of analysis common in political ecology (e.g., Blaikie 1985; Bryant and Bailey 1997), this paper attempts to show what resources were exploited, by whom and for whom and with what consequences, economically, politically and environmentally, over time. Thus, current socio-economic and environmental conditions are examined through the 'lens' of the historical trade in forest products trade within the context of shifting political relationships in Southeast Asia.

Unfortunately information about Laos, especially quantitative data, is rather scarce. As most trade statistics obtained proved inconsistent or incomplete over time, no quantitative analysis has been attempted here. Chinese dynastic annals provide the earliest written information on commerce and political relations in this region with details of trade missions and political alliances. Although rebellious tribes in Yunnan repeatedly interrupted contact between the ruling Han dynasty of China's central plains and the civilizations of South and Southeast Asia, trade became increasingly important as agricultural advances generated food surpluses and thus freed labor for other enterprise. Manufactured goods were exchanged for precious metals, strategic materials and exotic goods from neighboring lands. Interest grew in exploring foreign lands for new resources and in developing the political alliances that would foster conditions favorable to such trade, soon recognized to be a potential source of significant revenues. Records of China's early investigations of manufacture and commerce in neighboring lands are valuable sources of information about conditions in this region (Li 1979; Hu 1990).

Knowledge of the early Lao polity is similarly gleaned from the court chronicles of the first kingdoms. As with most court documents, these were designed to glorify and legitimize the ruling dynasty; they provide few details of economic activity or commercial relationships. The earliest of the Lao court chronicles date from 1422 to about 1500. For concrete information on Lao's international trade relations, we must turn to the documents of its trading partners and of early explorers to this region.

Pursuing commercial objectives similar to their Asian predecessors, European explorers (who arrived in Southeast Asia in the late 15th century) did not reach the Lao kingdoms for another century.[2] Two of the earliest visitors leaving the most detailed accounts of their brief visits were G. van Wuysthoff (1641-1642), the Dutch commercial envoy, and Father Marini, a Jesuit priest (1663). We do not read of Europeans in this area again, however, until the French survey expeditions of the mid-1800s (Mouhot 1868; Carne 1872; King 1995). Auguste Pavie, a French explorer and diplomat posted to the Lao capital in the latter half of the 19th century, was the first to translate the Lao Chronicles into a western language—French, with an English version appearing later. For this paper, historical documents in English, French or Dutch were reviewed when obtainable, while Asian language documents were examined in translation.

On the whole statistical records for the period before 1800 in Indochina are poor (Bulbeck *et al.* 1998). French colonial records of the late 19th and early 20th centuries are of variable usefulness given the tendency to aggregate much of the data over the whole of Indochina and the problem of shifting political boundaries. Post-World War II statistics are scarce given Laos' post-independence political isolation through the Vietnam War era. Even with regard to more current statistics, the World Bank (1997) notes that there is still no unbiased objective system for data collection on resources. Contemporary information on the trade in non-timber forest products, including endangered species, was developed with the assistance of the Department of Forest, Ministry of Agriculture and Rural Development in Vientiane and the World Conservation Union (IUCN), especially its Project on Non-Timber Forest Products. The author conducted fieldwork in both northern and southern Laos, visiting villages currently engaged in the cultivation and trade of various non-timber forest products. Further, visits were undertaken to markets in Laos and in neighboring Thailand, Vietnam, and Yunnan where many Laotian products are sold. Much of the trade in the informal sector, especially in illegal products, however, remains unrecorded.

Laos Today

To understand Laos today one needs to understand the physical context in which Lao society emerged. Accordingly this section provides a brief overview of the country's physical and biological characteristics before examining current socio-economic conditions. Subsequent sections will examine how the country arrived at its present state and discuss the implications of historical patterns of development.

The Physical Context

Lao People's Democratic Republic (PDR), or Laos, is the fourth smallest country of Southeast Asia (Table 3-1). The country consists of largely mountainous areas and highland plateaus with only a narrow strip of relatively flat land in the flood plain along its western border. The Mekong (the longest river in Southeast Asia) enters on the north and runs the entire length of the country; it forms the western border with Thailand. Landlocked Laos has 5,083 km of international borders abutting Thailand in the west, Vietnam on the east, China and Myanmar in the north, and Cambodia to the south. The highest peak, Phou Bia (2818 m a.s.l), dominates the Central Cordillera that runs the length of the southeastern border with Vietnam. This mountain range has proved a formidable but not impenetrable divide between these two countries. Stretching from 13° to 22° N latitude, the country enjoys a mostly tropical climate with a monsoon season (May to October) of almost daily precipitation. Rainfall averages from 1,500 mm in the lowlands to nearly twice that in the highlands. Maximum temperatures range between 25°C in January to 37°C in April with minimum temperatures sometimes falling to freezing in the higher mountain areas during the winter.

With only about four per cent of the land area under cultivation and almost half still under forest cover, Laos is very rich biologically. Most of the remaining forested areas consist of tropical montane rainforest stretching into subtropical broadleaf forests, which yield to coniferous species at higher elevations and in the far north. Tropical dry evergreen forests skirt the southern border. At the intersection of the Himalayan and Indo-Malayan biogeographical realms, the varied landscape with elevations ranging from 300 to nearly 3,000 m supports a high degree of endemism. In the last decade, four new mammal species, were discovered in the mountain range separating Vietnam and Laos (Schaller 1998). Unfortunately these discoveries are precisely in areas under greatest threat from increased forest exploitation and poaching by entrepreneurs, both from local villages as well as from neighboring countries (Le ca. 1995; Timmins 1995; Nash 1997; Donovan 1998).

Current Socio-economic Conditions

At 1998 estimates, the population of the Lao PDR now totals approximately 5 million, with an annual increase of more than two per cent; at this rate, it is projected to double in just over two decades. Currently, Lao PDR has the lowest population density in the region with an average of 22 individuals per square km; in parts of the country, this density is as low as eight persons per square km. Rates for life expectancy and literacy are significantly lower than those of most of its neighbors. About four-fifths of the population are rural farmers. The largely subsistence farming sector depends heavily on forest resources for many basic needs—mainly fuel, but also raw materials such as food, fiber, medicinal plants, and construction materials. It is estimated that more than 1.5 million people practice swidden cultivation. Government attempts to resettle people from hilly regions to roadside villages (more convenient for administering government programs) has met with limited success; many settlers eventually return to home areas.

Table 3-1. Basic indicators for selected areas of Southeast Asia, 1996

	Lao PDR	Vietnam	Yunnan[a]	Thailand
	(1)	(2)	(3)	(4)
Land Resources				
Total area (000 km^2)	236.8	331.7	394.0	513.1
Cropped area (000 km^2) (1996)	9.0	69.9	27.9	208.0
Forest & woodland area (000 km^2)	125.0	96.5	161.3	135.0
Population (1996)				
Total (millions)	4.8	75.3	40.0[d]	60.6
Annual growth rate, 1993–97 (%)	2.6	2.1	1.3	1.1
Density (persons per km^2)	20.0	227.0	101.0	118.0
Social Indicators (1996)				
Life expectancy at birth (years) M/F	51/54	66/70	63/65[b]	67/72
Safe water access (% of pop.) (1994–5)	41	43	-	81
Adult literacy rate (% of pop.)	57	94	37[b]	94
Economically Active Population (EAP)				
(% of total population)	48.8	50.5	53.0	53.2
Agricultural labor as % of EAP	77.1	68.9	79.0	38.1
National Accounts (1993–97)				
Per capita GNP, 1996 (USD)	400	290	166	2,960
Average annual real growth in GDP (%)	4.5	4.4	10.9[c]	2.2
Agriculture's share of GDP (%)	54.3	34.0	27.4	11.1

Sources: Columns (1), (2) and (4): ADB, 1998;
Columns (3): Statistical Bureau of Yunnan, 1993.

Notes:
a 1992 figures unless otherwise indicated
b 1990
c 1991–1992
d estimate

As a landlocked country, Lao's economy has always been highly dependent on neighboring countries in its relations with the larger international community. Accordingly, Laos has suffered with its neighbors in the economic setback that has hit the region over the past few years. Annual inflation rates have risen to 20 per cent or more. However, the country's so-called backwardness, that is, the economy's high degree of subsistence and self-sufficiency with regard to the production of basic goods, has also enabled the poorer segment of the population to weather well, and perhaps better than some neighbors, the shock of the 1997 Asian financial crisis.

The Lao currency, the *kip* (LAK), is essentially tied to the Thai currency (*baht*); its slide since the 1997 financial collapse has improved the competitive position of the country's exports. Although agricultural and service sector expansion boosting GNP growth rates to a respectable 7.2 per cent in 1997 (ADB 1998) seemed promising, the virtual stagnation since then has discouraged further investment. Historically Thailand has been the principal source of imports while recently Vietnam and China are the main destinations for exports, mainly raw materials (ADB 1998). Membership in ASEAN, achieved in July 1997, plus aspirations for most-favored-nation status with the United States and accession to the World Trade Organization (WTO), should help to keep the government focused on World Bank recommendations for fiscal discipline. The influence of its northern neighbor, Yunnan, appears to be increasing as China funnels an ever-increasing number of consumer goods southward, mainly in exchange for raw materials. Chinese consumer goods are now seen in almost every market in Laos. Financial assistance from Japan and ADB supports the improvement of the transportation system that links the country to the outside world. However, the inflow of private sector investment funds, most of which have come from Thailand, has been severely curtailed since the Asian financial crisis.

Laos' average annual per capita GNP of USD 400 and relatively high wage rates[3] (cf. Table 3-1) obscure the deep poverty in many rural areas. The disparity in wealth and living standards between rural and urban residents appears, already profound, is widening. An uncertain economic situation is made even more precarious by the increasing vagary of the weather. Of the more than 80 per cent of the population who live in the uplands, most are ethnic minorities with scarcely any contact with the capital city. Numerous families in the rural areas experience shortages of rice for three or more months per year. Selling livestock or shouldering debt to obtain cash are widespread. The collection and sale of non-timber forest products (e.g., bamboo shoots, mushrooms, and wild animals) is an important source of household income in rural areas, comprising from 30 to 70 per cent in some cases (Laurent 1990; Foppes *et al.* 1997; Nash 1997; Suksavang 1997; Enfield *et al.* 1998).

By all formal measures Laos is one of the poorest countries in the region and in the top ten of the poorest countries worldwide. As illustrated in Table 3-1, social and economic indicators lag significantly behind those of neighboring states. Rural areas, for centuries the source of valuable forest products, today demonstrate little evidence of the wealth generated by the exploitation of these resources. A closer look at the development of the trade in this region over time may reveal some explanation of these disparities.

Lao History:
From International Trade to Isolation and Back

Laos' development history is essentially characterized by a sequence of power struggles over resources, namely land and people. In pre-modern times as paddy-based agricultural societies expanded and spread through the lowlands of mainland Southeast Asia, displaced groups sought refuge in adjacent hills, where they established their own polities (Provencher 1975; Wolf 1982). Given the uplands' biological richness and geographical isolation, people who settled here could be largely self-sufficient. The system of swidden agriculture complemented by forest exploitation that evolved remains the most successful form of livelihood for most upland populations in the region.

Political relationships among neighboring groups in this region were essentially hierarchical. In the classic model, a regional ruler presided over several local rulers who in turn presided over other smaller chiefs. The smaller states paid 'tribute' to the larger states as an indication of their deference, loyalty and support. The tributary states were still considered separate kingdoms with each maintaining its own network of hierarchical relationships, supported by court and administrative structures, financial systems, tax collections, armies, and judicial systems. In essence, then, local rulers recognized the potential benefits of paying up as opposed to being the object of pillage and plunder. For his part the regional ruler was expected to police major communications and trade routes. He was also supposed to protect villages and caravans from bandits, but this service mainly benefited the wealthier, trading classes, that is, the nobility and the merchants. These early tributary relationships formed the basis of political, social, and economic relations that influenced the subsequent development of nation states in this region.

Contested Sovereignty
From the second to the 12th centuries a succession of powerful Khmer kingdoms from Funan to later Chenla and Angkor developed in the lowlands of what is now Cambodia. Important in early trade with the west and China, they exerted wide-ranging political influence, from Laos in the north down to the South China Sea and through Thailand in the west across to the Malayan Peninsula. Over this period of time, many separate polities developed in the montane regions. The first unified Lao kingdom, Lan Xang, was founded in 1353 by a prince trained at Angkor (Lebar and Suddard 1960). At its zenith the Lao territory included parts of what is now

Yunnan, western Vietnam, northeastern Thailand and Cambodia (Berval 1959:407). Internal rivalries persisted among the nobility, however, and, as Table 3-2 illustrates, Lao history has been marked by a long series of struggles, both internal and external, as ambitious kin and neighboring states repeatedly vied to take control of this mountainous area and its rich resources. At one time or another the Lao kingdoms were tributaries of Vietnam, Thailand, Burma or Cambodia, often owing several overlords at any one time (Coedes 1967; Wolters 1982; Stuart-Fox 1997). Such multiple, overlapping allegiances and ambiguous frontiers meant that the sovereignty of the individual tributary states was always in question (Reinach 1901). As a result, disputes over territory and intrigues over access and control of resources were a near-constant distraction for the ruling elite.

In the mid 19th century, France, from a base in Cochin China, spread its influence and control northward, absorbing Cambodia as a protectorate in 1864 and subsequently Tonking and Annam two decades later. French explorers repeatedly tried to find an access route to central China, perceived as an area of enormous riches, via the Mekong River. The Siamese used the incursion of groups fleeing south from China's Taiping Rebellion (1850–1864) as the excuse to occupy its northern border states, including the Lao kingdoms. Only a few decades later, in 1893, France relieved Thailand of these acquisitions (east of the Mekong), in effect, absorbing Laos as one of its Indochinese protectorates. Laos was to remain under French administration essentially for more than a half century, interrupted only briefly by Japanese occupation in 1945. After World War II, France again assumed control but rebellion was fermenting among their colonial subjects. Following several abortive attempts at independence and the spillover of the Vietnam War into Laos, local communists finally seized power in 1975 and established an independent government. Although this fledgling government was initially buoyed by the Soviet Union and other Eastern European communist states, support soon faded as they themselves ran into political and economic trouble. Despite subsequently warming of relations with its communist brothers to the north and east, Lao rulers remain wary of these historical enemies and economically stronger neighbors as well as its capitalist neighbors to the south.

Development of Trade and Commerce
Trade has played a pivotal role in the development of Southeast Asia, which historically was both a primary source of raw materials as well as an important entrêpot between western and eastern societies. The role of

Table 3-2. Significant milestones in Lao history

Year	Event
1353	Founding of Lao kingdom of Lan Xang
1358	War against the Siamese
1477–1479	Vietnamese incursions repelled
1563	Vientiane becomes capital in Kingdom's golden period of prosperity
1575	After a decade of intermittent war Lan Xang becomes vassal state of Burma
1591	Lao throne regains independence
1641–1642	First Europeans to leave records arrive in Vientiane
1707–1713	Internal struggles result in Lan Xang division into three kingdoms
1779	Northern Lao kingdoms become tributaries of Siam
1820–1840	Repeated waves of emigration from China
1861	French explorer Henri Mouhot arrives and later dies in Luang Prabang
1864	French annex Cochin China taking Cambodia as a protectorate
1866–1868	French send exploration mission up the Mekong
1884	French annex Annam and Tongking
1887	Auguste Pavie, first French vice-consul, arrives in Luang Phrabang
1893	Siam cedes Lao territories to France after repeated clashes
1907	Franco-Siamese treaty establishes present frontiers of Laos
1945	Japanese invade and 6 months later surrender
1946	French reoccupy Laos
1953	Laos reorganized as fully sovereign state within French Indochina Union
1955	Kingdom of Lao admitted to United Nations; Lao People's Party formed
1964–1973	As war rages in neighboring Vietnam, U.S. bombs communist targets in Laos
1975	Lao People's Democratic Republic proclaimed
1978–1979	Agricultural cooperatives program launched and abandoned
1986	Fourth Congress of the Lao People's Revolutionary Party endorses New Economic Mechanism introducing market economic principles
1990	Chinese Premier Li Peng visits marking improving relations with China
1993	Further economic reforms and environmental protection law passed
1994	First bridge across Mekong connects rail and road system of Laos and Thailand
1994	Renovation on Highway 13 to link Saigon and Kunming

Sources: Berval 1959, Lebar and Suddard 1960, Stuart-Fox 1997

China in this history is critical. As early as the 2nd century BC China conducted regular trade with South Asia and Burma through trade routes in Yunnan. In 221 BC the ruling dynasty ordered an expeditionary force of half a million strong to march south in search of precious goods, such as rhinoceros horn, ivory, jade and pearls (Li 1979). By 111 BC Tonking (northern Vietnam) had been annexed as a protectorate of the Chinese empire and tribes on the frontier were required to pay tribute to a ruling administration based in Canton (Wheatley 1979). Trade routes stretched south through Yunnan along the major river systems—the Irrawaddy and the Salween into Burma, the Mekong through Laos and Thailand, and the Red River in Tonking (Purcell 1951). With overland routes through central Asia regularly plagued by bandits, in 226–230 AD China sent emissaries to Funan (Cambodia) seeking sea routes to the rich resources of Southeast Asia.

As the richest and most administratively developed state of the period, China was able to raise and equip an army that enabled it to control the international trade routes that delivered foreign goods to its doors and persuade neighboring polities to pledge their allegiance to it. Obligatory gifts from these vassal states, the ostensible object of which was to cement political alliances, were delivered on a regular basis, usually annually or triennially (Winichakul 1994). Although the principal payment would be made in gold or silver, a tributary state was also obliged to send labor, goods or other supplies as requested. Tributes from China's vassal states included forest products. These goods were valued for their utility or novelty and consequent prestige of ownership and, significantly, for the large profits they could earn for the crown in subsequent resale. Participants in early tributary missions soon learned how to profit from these ventures. Government officials, merchants, sea captains, and sailors—all became involved in trading (Ho 1935, Chang, P. 1991, Gunn 1999). Ships' manifests indicate as much as 13 to 17 per cent of total cargo weight in ships bound for China were on taken consignment or were property of the crew (Cushman 1993, App. B).

Trade between China and South and Southeast Asia was instrumental in the spread of Buddhism. In addition to religious and philosophical texts came the pharmacopoeia of the Buddhist homeland, thus generating a demand for many plant and animal products from South and Southeast Asia. Botanists estimate that 75 per cent of the plants native to Southeast Asia and now grown in China arrived before 1000 AD (Hu 1990:491). By the 12th and 13th centuries, aromatics, spices and natural products used for medicinal preparations were the most important items, in terms of value and volume, in trade with China (Wheatley 1959). As shown in Table 3-3, such products constituted nearly half of Southeast Asian exports to China (Chang, P. 1991).

Archaeological evidence indicates that the early kingdoms of Southeast Asia were probably relatively sophisticated agricultural societies with extensive areas of wetland rice under cultivation. It is unlikely that these

Table 3-3. Tribute goods and exports of early major trading centers in Southeast Asia

KINGDOM AND ERA	SELECTED EXPORT ITEMS*	SOURCE
Funan ca. 2nd to mid-6th century	Eaglewood, gold, silver, aromatics, pearls	Wang 1958; Li 1979; Provencher 1975
Angkor [Cambodia] ca. 9th to mid-15th century	Ivory, eaglewood, beeswax, kingfisher skins, damar resin, incense, sapanwood damar resin, benzoin, raw silk, rhino horn, birds' nests, sandalwood, gambodge, cardamom, lakawood, rattan, tiger and leopard pelts, rhino skins, ape skins, warblers	Hirth and Rockhill 1967; S. Chang 1991; Wicks 1992
Champa [Vietnam] Ca. 2nd to 15th century	Ebony, eaglewood, rhino horn, camphor, gold, aromatics, ivory, elephants, monkeys, cinnamon, lac, sapanwood, sandalwood, birds' nests, pepper, cardamom, lakawood, kingfisher skins, beeswax, rattan, tiger and leopard pelts, rhino skins, ape skins warblers, tortoise & turtle shell, bamboo	Ma Huan 1433; Fillastre 1905; Manguin 1972; S. Chang 1991; Wicks 1992; Momoki 1998
Ayutthaya [Thailand] mid-14th to mid-18th century	Eaglewood, lakawood, rosewood, sapanwood, cloth, gold, jewels, benzoin, gomma lacca, incense, elephants, tortoise, black pepper, monkey, deer hides, lead, tin, gems, rhino horn, pepper, ivory, benzoin, birds' nests, sandalwood, eaglewood, cardamom, sticklac, dragon's blood, cutch, gambodge, cardamom, red sandalwood, bamboo, kingfisher skins, beeswax, sapanwood, rhino skins, warblers	Ma Huan 1433; Marini 1663; S. Chang 1991; Wicks 1992; Viraphol 1997
Annam [Vietnam] 12th to 19th century	Black pepper, areca palm, nutmeg, sapanwood, ebony, tropical cypress, cardamom, rhino horn, birds' nests, deer tendons, fragrant snails, tortoise shells, ivory, iron, zinc, medicines, cinnamon, pearls, eaglewood, red sandalwood, beeswax, rattan, parrots	Hirth and Rockhill 1967; Woodside 1995
Tongking 14th to 16th century	Eaglewood, rhino horn, cardamom, birds' nests, tortoise shell, benzoin, lakawood, deer horns, ivory, kingfisher skins, bamboos, beeswax, sapanwood, rattan, ebony, snake skins, rhino skins, ape skins, deer skins, deer meat, warblers	S. Chang 1991

*No significance in order presented. Cultivated products largely excluded.

lowland societies themselves produced the many exotic forest products common on export lists. Moreover, expansion of agriculture and a growing population was transforming the landscape in Southeast Asia as it had southern China[4]. The forest products featured prominently on export rosters of the major trading ports of mainland Southeast Asia were collected from tributary states and from minority ethnic groups in their hinterlands (Crawfurd 1830, Wicks 1992, Reid 1993, Breazeale 1999). Powerful states, such as Annam or Ayutthaya, could have a dozen or more tributary states, such as Laos. Ayutthaya's fame as a trading center has been attributed to its success in organizing regular supplies from its network of sources in the interior (Breazeale 1999:20).

Through initiatives of Middle East merchants, products from Southeast Asia had long been familiar to consumers in Europe. But it was not until the 15th century that western European traders made serious attempts to capture a share of the east's immensely profitable spice trade. The Portuguese, who first reached the Malabar coast in 1498, established trade relations with Cochin China between 1525–1530 (Manguin 1972). The Spanish, Dutch, English, and French soon followed. As with previous merchants, these newcomers also sought products of high value-to-weight ratios, such as spices, pearls and other natural products unique to this region. In return, they offered gold and silver from the Americas, firearms (popular with ambitious local rulers), and, later, manufactured goods. By the mid-19th century, the major European trading nations had largely displaced Asian competition in the sea trade. In reality, though, given the strength of the Asian tributary system, these Western interlopers were reduced to 'mere, albeit highly enriched, middlemen' (Gunn 1999:176). Having recognized the benefits of peaceful coexistence, by the beginning of the 19th century they had settled into a wary peace with their contemporaries and retired to their separate spheres of influence[5] to reap the rich rewards of colonial administration.

For Laos, the Mekong River was the major artery connecting it to the outside world. Most of the goods moving into or out of this region traveled via the Mekong, generally during the high-water season of July to December. The river was not passable year-round, and rapids at several points necessitated portage of goods overland even in good times, putting trading parties at risk from wild animals as well as bandits. Whenever the route to the south became impractical, porters, if not ponies or mules, carried loads east through the high mountain passes of the Central Range to Vietnamese ports on the South China Sea. Products continued to flow regularly northward through Yunnan in the able hands of Chinese Muslim traders (Hill 1998). The 19th century development of the rail and road systems across the Korat plateau in northeastern Thailand facilitated the flow of goods southward. It was then also financially feasible to transport bulkier, lower value commodities from collection points in Laos to Bangkok (S. Chang 1991).

By the 18th and 19th centuries, trade and commerce had become a main source of government revenues for the major trading states of Southeast Asia (Wheatley 1959, Viraphol 1977, Manarungsan 1989, Reid 1993). Over time, tributary payments and formal exchanges of gifts between rulers had differentiated into 'crown trade' and 'private trade'; several elements of society participated and a rich and powerful merchant class developed as intermediaries in this trade (Wheatley 1959, Chang, P. 1991). Trade became an ever more important source of revenue as successive administrations monopolized specific products as exclusive rights of the crown (following the ancient tradition of the common government salt monopoly). Over time the number of goods listed as a royal monopoly expanded along with the number of taxes and fees on trade. By the beginning of the 19th century, almost all the internationally traded goods in Thailand—including eaglewood (a highly prized aromatic), ivory, gambodge and pepper—were royal monopolies. By the latter part of the Ayutthaya period, overseas trade revenues amounted to as much as 36 per cent of the Siamese state's total cash income (Breazeale 1999). Similarly, records show that eaglewood was one of the early royal monopolies in Vietnam (Fillastre 1905).

From the mid-19th century to the beginning of World War II, colonial administrations adopted an orientation that was to have a significant affect on the economies and societies of Southeast Asia. Colonial policy overseas was in essence driven by industrial policy at home. European countries looking for raw materials and markets turned to their colonies to meet these objectives. Tropical plants were transported from one continent to another to establish plantations and enhance production. Thus, rubber came to be prominent in the exports of Malaya and the East Indies; production of sugar and coffee was also promoted. New plantations, often established with foreign funds, were worked by immigrants (imported workers). French investment in irrigation works and other infrastructure contributed to increased productivity in rice production. Rice surpluses from Indochina were exported around the region to feed populations now focused on the production of cash crops. Thus a thriving intra- as well as interregional trade developed.

Between 1918 and 1930 the French colonial administration in Indochina launched a comprehensive program of development in an attempt to establish the preconditions for capital expansion and improve conditions of production. Extensive investments were made in infrastructure development. During 1926–1929 foreign investment (mainly French) increased especially in the mining sector. In 1928, a State Forestry Service was established in Laos to develop timber production, especially of teak (Gunn 1990: 25; Potter, this volume). In 1929 a glut in the world grain market caused a dramatic drop in rice prices. The worldwide economic depression that had hit the West was also affecting the prices for other commodities in the region. The effects cascaded through the trade-linked colonies. As prices for their export commodities fell, the French, among others, resorted to retrenchment and restriction schemes in an effort to shore up prices for their commodities (Latham 1997). Statistical records indicate that while forest products exports from the French colonies held almost steady over the period 1927–1936, agricultural products increased slightly with the majority headed to France (Service de la Statistique Generale 1927, Bureau de la Statistique Generale 1937).

At the end of World War II, Laos had barely 20 industrial enterprises: half were rice mills and the remainder sawmills, paper mills, tin mines, rubber or coffee plantations (Gunn 1990:28). Nearly three decades were to pass before Laos gained real independence, with a communist government in control. Subsequent policies of political and economic isolation from former trading partners (non-communist) brought the already vulnerable economy to near-collapse by the mid-1980s. Following the prevailing trend in the region, Lao leaders in 1986 began to introduce economic reforms, which included opening to the West, a process that accelerated after the Soviet Union's 1989 collapse. Political and economic reform continues. Not only has Laos been welcomed back into the larger commercial community, but its resource wealth is greatly coveted by neighbors. Recent investments to upgrade the transportation network have improved communications as well as trade links to both neighboring countries and the world beyond.

Patterns in Trade and Development Over Time

The long history of trade in Southeast Asia and the region's dual position as both entrêpot and source of supply highlights the key role played by international commerce in the economic, social and political development of this area. While most historical analyses have focused on the primary power centers in this drama, i.e., the major port polities, this paper focuses on the interior, the major source of supply. Notwithstanding the dearth of quantitative data on specific products or particular areas, such as Laos, it is useful to examine the pattern of trade over time in an attempt to understand its probable impact, especially in upland areas. Accordingly the following section explores the composition, direction, pace and participants in the forest products trade in an effort to learn more about its influence.

The Composition of Trade

Throughout history market dominance has shifted from port to port across the region. However, many of the same products, some relatively obscure in western experience, have continued to play an important role as trade items. Comparing Tables 3-3 and 3-4 with Tables 3-5, 3-6, 3-7, and 3-8, we see that the roster of exports, many of which would have come from Laos, include many of the same items. These were products of forests and forest fallows as well as cultivated areas. Cardamom, for instance, is a semi-domesticated crop is now harvested from cultivated fields as well as forests. First in volume at the beginning of the century, it remains the export leader with volumes having increased almost tenfold.

For other forest species, such as spices, domestication has increased availability, thus lowering prices and demoting these products from their once luxury status. In recent years, there has been a revival of interest in natural products. For example, indigo, a traditional dye made from a plant cultivated across the region and largely replaced by synthetic dyes in the 19th century, has enjoyed a comeback. Similarly interest grows globally in traditional Chinese medicines, which in turn contributes to the strong demand for both wild plant and animal products from Southeast Asia, including Laos (Donovan 1999). For many other products, e.g., eaglewood, for which there are neither acceptable substitutes nor successful cultivation regimes, increased demand has led to overexploitation, scarcity, and higher prices, and in some areas of its range the species is now threatened with extinction (Barden *et al.* 2000).

The composition of trade reflects not only consumer preferences, but also product availability. The trade lists noted in the earlier Tables 3-3 and 3-4 indicate that these products were readily available for several centuries. Indeed, except for Cochin China (largely the delta area in southern Vietnam), statistics show that forest cover remained relatively good throughout Southeast Asia well into the 20th century. French colonial records of 1933 estimate that over the whole of Indochina there was a total

Table 3-4. Total exports from lower Laos, 1899

Article	Average Price per Picul (piastres)	Volume (kg)	Total Value (piastres)
Agricultural Products			
Betel nut	9–10	180	29
Cotton	3–4.50	240	16
Ginger	3	127	7
Pepper	9	240	36
Tobacco	18–24	168	60
Sub-total			148
Livestock Products			
Buffalo hides	8–12	34,560	6,422
Cow hides	10–14	25,020	5,035
Horn of cow and buffalo	7–12	25,480	3,640
Sub-total			15,097
Forest Plant Products			
Cardamom	12–23	267,060	84,089
Chinese medicine	15	540	135
Chinese nettles	7–20	42,390	10,608
Palm fiber	3	360	18
Sub-total			94,850
Forest Wildlife Products			
Beeswax	60–80	6,150	7,805
Carapace of turtle	9	120	18
Deer horn	9–15	11,640	2,577
Deer horn (soft)	200	-	112
Deerskins	9–20	38,280	9,305
Elephant bones	3	126	7
Elephant fat	5	-	1
Elephant ivory	180–400	360	2,066
Monkey skins	20	90	30
Pangolin scales	12	30	6
Pangolin skins	11–12	700	136
Rhinoceros horn	55 per kg	38.5	2,118
Rhinoceros skin	17	330	93
Shellac	9–11	14,660	2,439
Tiger bones	15–18	130	35
Tiger skins	4–7 per item	1	7
Sub-total			26,755
Minerals			
Gold dust	12.70 per ticul	21 ticuls	268
Lime	0.50–0.60	9,000	82
Tin	36	4,500	2,700
Sub-total			3,050
Manufactured Products			
Bamboo and wickerwork	1 per item	40 baskets	40
Canoes	By size	-	22,076
Cotton goods	0.10 per m	80 m	8
Silk thread	120	180	360
Sub-total			22,484
Total			162,384

Source: Derived from Reinach 1901

Table 3-5. Major forest plant products entering international trade from Lao PDR, 1995/1996

Trade name	Scientific name	Commercial Portion(s)	Official export volume 1995–1996 (MT)
Bamboo	(various)	Stems	#
Rattan	Daemonorops spp., Calamus spp.	Stems	2,600
Cardamom	Amomum spp.	Fruits	2,200
Broom grass	Thysanolaena maxima	Flower	420
Paper mulberry	Broussonetia papyrifera	Bark	370
Bong bark*	Nothaphoebe umbelliflora	Bark	300
Pobit, tut-tieng*	Helicteres isora	Bark	200
Chandai*	Dracaena cambodiana	Wood	130
Vang dang*	Combretum extensum	Stem	130
Benzoin	Styrax tonkinensis	Resin	53
Ferns*	Heminthostachys sp.	Roots	50
Rattan	Daemonorops sp.	Fruits	20
Eaglewood	Aquilaria crassna	Wood	18
Orchids	(various)	Entire	12
Bamboo	(various)	Shoots	10

Source: Department of Forest, Government of Lao PDR, 1995–96

Notes:
* Plants used for medicinal preparations.
# Total commercial volume was 5 million tons. The breakdown between exports and domestic consumption is not known.

of approximately 42,400 million ha of forested land area. Just over half of these were in Laos (Bureau de la Statistique Generale 1937:124). The spread of plantation crops like tea, coffee, rubber, and oil palm started a trend that continues to be significant in reducing forest cover.

Price trends, when such information is available, may be a better indicator of product availability, as rising prices reflect increasing scarcity relative to demand. The high price observed for many wildlife products indicate not only their significant decline in numbers, however, but also their protected status. As a result of threats to population viability of species like tigers, elephants and rhinoceros,[6] most national governments in this region have passed conservation laws prohibiting hunting and trade in these species (Donovan 1998). Nevertheless, even a decade ago as many as 80 per cent of villagers interviewed in one survey expressed concern that wildlife populations were decreasing (Salter 1993). As more recent interviews with inhabitants in other Lao villages attest, villagers recognize that they are losing valuable forest assets and blame the pressure of outside markets as well as the demands of a growing local population (Foppes and Ketphanh 2000). Species disappearance is due, however, to habitat reduction as well as hunting. Expansion of cultivated areas and overhunting have reduced the number of elephants and rhinoceros to one or two isolated groups in very remote areas.

Despite the host of signatories to the international agreement restricting trade in endangered species (CITES), customs interception records and various field reports indicate that the trade in rare animals and their products continues. The species listed in Tables 3-6 and 3-7 are those whose trade is prohibited or restricted according to national law in

Table 3-6. CITES-listed wildlife products entering international trade from Lao PDR, 1998

Trade Name	Scientific Name	Commercial Portion(s)	CITES Status*
Mammals			
Elephant	*Elephas maximus*	Ivory	I
Javan rhinoceros	*Rhinoceros sondiacus*	Horn	I
Tiger	*Panthera tigris*	Pelt, bones	I
Leopard	*Panthera pardus*	Skin	I
Clouded leopard	*Neofelis nebulosa*	Skin	II
Black bear	*Selenarctos thibetanus*	Gall bladder	I
Kouprey	*Bos sauveli*	Horn	I
Chinese pangolin	*Manis pentadactyla*	Scales	II
Douc langur	*Pygathrix nemaeus*	Liver	I
Francois langur	*Presbitys francoisi*	Liver	I
Silver langur	*Presbytis cristatus*	Liver	I
Gibbon	*Hylobates* spp.	Liver	I
Birds			
Thick bill pigeon	*Treron sieboldii*	Liver	II
Spotted dove	*Streptopelia chinensis*	Liver	III
Grey peacock	*Prolypectron bicakaratum*	Feathers; Liver	I
Peafowl	*Pavo mutiacus*	Feathers; Liver	I
Hill myna	*Gracula religiosa*	Liver	I
Reptiles			
Box terrapin	*Cuora amboiensis*	Liver	I
Rock python	*Python molurus*	Skin	I
Reticulated python	*Python reticulatus*	Skin	I
Siamese crocodile	*Crocodilus siamensis*	Skin	I

Sources: Wildlife Conservation Project, Dept. of Forest, Government of Lao P.D.R., 1998

*Convention on International Trade in Endangered Species categories: I: prohibited; II: protected; III: noted.

accordance with the CITES agreement. With few alternative sources of income, plus the high prices offered for this contraband, village hunters are willing to risk the legal and ecological consequences of the trade (MacDonald 2001). National and international conservation laws do little to stop this lucrative business. Given the paucity and inconsistency of data on Lao exports, especially in non-timber forest products, analysis of the economic and environmental effects of this trade is limited. Given that most of the trade in wildlife and a number of wild plants is informal and much of it now illegal, there is no record of these transactions. However, expert surveys conducted in rural villages and at border checkpoints support the widespread opinion that there is an unmistakable decline in almost all wildlife species. Farmers, former hunters, interviewed in Vietnam readily admit that there are almost no animals left in the forest apart from rats and the passing bird.

An analysis of commercial statistics suggests patterns of both change and continuity in trade. The high-value, low-weight products are no longer dominant as improved roads and better vehicles make the export of heavier, bulkier, lower value products, such as timber and coffee, financially attractive. Thus, the product mix is influenced not only by overseas and local markets, but also by the willingness of investors to invest in production or extraction. Recently timber has become a significant source of government revenues as various companies, primarily foreign, have invested in the infrastructure necessary to exploit and develop these resources. But product availability can also be to some extent supply-driven. The French colonial administration used a cash-only tax policy to force hill farmers to bring their goods to market to exchange for cash and thus enter the cash economy. This departed from the former system of paying taxes, or tribute, in 'kind', i.e., goods. In today's economy local hill farmers need cash even more, to pay not only taxes, but fees for the health clinic, medicine, and school, among other charges, or to purchase the numerous consumer goods flooding in from Thailand and China.

In sum, we see that over time the composition of trade has not so much shifted from one set of goods to another, as restructured and expanded. New additions to the product list have been made possible due to technological developments, which enable the exploitation of a wider array of forest products. Meanwhile many traditional products remain but have been relegated to the informal sector, or even black market trade. This is highlighted by the discrepancies observed between the official list of exports and those of customs interceptions and the environmental agencies (Tables 3-6, 3-7 and 3-8).

Table 3-7. Recorded export and destination of CITES-listed wildlife species from Lao DPR, 1983–1990

Species	Quantity	Main Destination
Mammals		
Primates (live)		
Hylobates lar	21	USSR
Macaca arctoides	20	Japan
Macaca fascicularis	120	U.S.A.
Macaca mulatta	350	Japan, U.S.A.
Macaca nemestrina	50	Japan, U.S.A.
Nycticebus coucang	420	Japan, U.K., U.S.A.
Nycticebus pygmaeus	12	Sweden
Pygathrix nemaeus	5	U.K.
Pangolin Manis javanica (skin)	7,000	Japan
Elephant Elephas maximus (tusk)	95	U.S.A.
Others		
Aonyx cinerea	73	Japan, U.S.A.
Felis chaus	8	Japan
Ratufa bicolor	50	Japan, U.K., U.S.A.
Birds		
Birds of Prey		
Accipiter badius	36	Japan
Elanus caeruleus	160	
Spilornis cheela	9	
Spizaetus nipalensis	110	
Microhierax caerulescens	122	
Tyto alba	85	
Hornbills Buceros bicornis	130	Japan, Singapore, U.S.A.
Reptiles		
Snakes		
Python molurus bivittatus	2,600	U.S.A.
Python reticulatus	350	U.S.A., Belgium
Ptyas mucosus	10,000	Spain
Lizard Varanus salvator	30	Belgium
Tortoises/Turtles (live)		
Manouria emys	40	Japan
Testudo horsfieldi	90	Japan

Source: Derived from Nash and Broad (1993) and Salter (1993)

The Direction of Trade

As Karl Polanyi (1975) observed, from an institutional perspective trade is simply "a method of acquiring goods that are not available on the spot." Thus trade (and previously tributary relationships) in Southeast Asia

Table 3-8. Summary of non-timber forest products exports to neighboring countries during first six months of FY 1996–1997

Trade Name	Scientific Name	Units	Volume	Value (000 LAK)
Exports to China				
Cardamom	*Amomum* sp.	mt	12.0	58,326
Bong bark	*Nothaphoebe umbelliflora*	mt	136.0	46,061
Rattan	*Daemonorops* sp., *Calamus* sp.	mt	10.0	5,992
Teak seed	*Tectona grandis*	mt	7.0	2,100
Chandai	*Dracaena cambodiana*	mt	3.0	1,110
Benzoin	*Styrax tonkinensis*	mt	0.8	800
Carapace of turtle		kg	50.0	40
Sub-total				124,926
Exports to Thailand				
Broom grass	*Thysanolaena maxima*	mt	54.0	20,760
Makken	*Zanthoxylum nitidum*	mt	8.0	4,200
Sugar palm fruits	*Arenga pinnata*	mt	3.5	1,617
Kenaf	*Hibiscus cannabinus*	mt	2.5	1,250
Eaglewood	*Aquilaria* sp.	kg	41.5	1,129
Sub-total				28,956
Total				153,882

Source: Ministry of Commerce, Government of Lao P.D.R., 1997.

Note: LAK in the international symbol for the Lao currency, kip; in August 1997 1300 kip = one USD

moved natural resources from areas in which they were relatively plentiful (e.g., Laos) to areas in which they were not (or no longer, e.g., China). Facilitating this exchange were the economic surpluses and technological advances generated in certain societies. This is still the case today.

From historical records available, it appears that the structure of long-distance trade in Laos has followed the generally recognized pattern described earlier. Products accumulated at a series of collection points, were sorted by quality or some other distinguishing factor, then channeled to ever-higher levels of aggregation. Finally, they were directed to the market with the highest potential reward (Reinach 1901, Berval 1959, Skinner 1964). With the development of trade from tribute relationships, the commercial network exhibited a hierarchical pattern in part reflective of political relationships. From Laos products moved in four directions: north to Yunnan, east to Annam, southwest to Thailand, and south to

Cambodia and Cochin China. Indeed, along the length of the country goods would move downstream accumulating at collection points, later to become major towns and cities, along the Mekong, where most commercial affairs were organized by resident Chinese or Vietnamese businessmen with links to larger trading networks.

With water both a cheaper and more secure method of transport, the major river systems of this region have been the main commercial arteries in this area since earliest times (Noone 1954, Elson 1997). Topography, though, has not been the only directional determinant of trade flow. Periodically natural disasters, such as landslides or floods, or political instability accompanied by increased lawlessness, would force merchants to abandon one or more routes. Banditry, warfare, and epidemics were also common disruptions (Elson 1997:21). If natural or political disasters failed to alter trade routes, institutional factors could take their toll, literally. Punishing taxes, excessive customs duties or other burdensome regulations would prompt traders to find new outlets. In addition to the regular fees, 'special' payments demanded by officials at all levels, from head mandarin down to clerks who prepared the letters of safe passage, were particularly onerous to merchants facing increasingly stiff competition in the busy international trade environment of the 19th century (White 1823).

Over time advances in technology and investment in infrastructure served to reduce the physical impediments to trade. The development of road and rail links across the Korat plateau in northeastern Thailand in the latter part of the 19th century as well as a more favorable tax structure encouraged Lao exporters to redirect their goods to Bangkok as opposed to Phnom Penh and Saigon. Advances in technology made the communication of market information, especially prices, much easier and more efficient for those with access to this technology. While the first bridge over the Mekong linking the Thai and Lao road and rail systems was not opened until 1994, the upgrading and extension of Route 13 will eventually link Saigon and Kunming. The impact of this new highway will be considerable, both for trade patterns and environmental conditions. Extending the length of the country and running through several conservation areas, this new corridor will provide forest entrepreneurs with direct access to the largest and fastest growing market in Asia, namely China.[7]

The Pace of Trade

Advances in technology have probably most significantly influenced the pace of trade over time. Early sailing ships in the long-distance sea trade were dependent on wind and water currents and therefore limited to seasonal travel. Moreover, many of the traditional vessels of the early 19th century were still relatively small; a Chinese junk carried a cargo of approximately 5000–8000 piculs while a four-masted schooner could carry an additional 50 per cent (Cushman 1993: App. B). In Malacca in the 16th century, there were at any one time probably more than 2000 vessels lying at anchor (Gunn 1999:7). In addition to weather, war and disease could affect the level and pace of trade. Epidemics affected not only the availability of collectors, porters and pack animals, and seamen, but also the willingness of seamen to enter an affected port.

In many agricultural societies in Southeast Asia, forest products collection is integrated with agricultural production (Ichikawa, this volume). To what extent labor was redirected from agriculture subsistence activities to the collection of tributary products is unknown. Although there are reports of specially organized collecting expeditions, e.g., for eaglewood, most products were probably collected over an extended period of time. Except for live animals, most products in the early long-distance trade were neither seasonable nor readily perishable. Merchants collected and stockpiled products for shipment, which before the late 1800s occurred only a few times each year.

Technological advances of the late 19th century marked a turning point in trade relations in Southeast Asia. Advances in navigation, nautical design and engineering (e.g., the steam engine) as well as the efficient management systems and the development of financial services, contributed to European success in maritime commerce. Recent advances in communication technology portend to have the greatest impact on trade from formerly remote regions such as Laos. Merchants using cellular phones linked to satellite systems now communicate directly with clients while negotiating with farmer-collectors in the field. Up until the latter part of the 20th century, the pattern of boom and bust in many non-timber forest products markets, the intermittent nature of tribute payments, and limited shipping schedules resulted in forest products collection being distributed over a longer period. This pattern of exploitation gave heavily tapped resources an opportunity to recover. Given the extreme connectedness of today's markets, emphasized by the development of transport and communication technologies, what may have once been intermittent stress on resources has now become almost constant pressure.

The Beneficiaries of Trade

To understand the beneficiaries of trade, one must examine the entire length of the trading chain, from the producer, or the collector, through various intermediaries to consumer. This includes not only those who handle the good, such as the traders, but those who collect tax or fees on commercial activity, such as ruling elites who derive their income from these sources.

In the early days of tribute and trade most primary producers were probably involuntary participants in this activity. Residents of the trading nations as well as the tributary states were required to pay taxes in kind and specifically in those goods that the crown could export (Viraphol 1977, Reid 1993). Racially differentiated tax systems in Indochina weighed disproportionately on the upland population who delivered their required payment in forest products (Gunn 1990:55; Li 1998:136). Those who wished to trade in forest products were required to deliver goods regulated by royal monopoly to government agents who paid suppliers a fixed price. For other products merchants were required to deliver a given proportion of their receipts to the crown. Such traditional systems were often retained by subsequent colonial administrators. Arguably, many aspects of forest administration in this region today reflect these age-old cultural views regarding proprietorship of forests and the role of forest resources (and one could add forest-dependent peoples) vis-à-vis the national treasury.

Until recently, indigenous groups probably were the sole extractors of most forest products in Laos. Research across the region has shown that the producer is very likely the individual or group to benefit least from trade. Two common characteristics of collectors' condition, inadequate recompense and a lack of information, have a significant effect not only on economic condition but resource management strategy and therefore environmental conditions. Still today suppliers in Laos operate in a situation of almost no market information and little if any technological support (Enfield *et al.* 1998).

The main beneficiaries of trade are commonly the intermediaries. This is true not only in Laos but also throughout the region (Padoch 1992, Enfield *et al.* 1998). Many of the insurers, merchant bankers, shippers and agents in Bangkok, London and Hamburg today (e.g., Lloyds of London) got their start in these headier markets of former years. Table 3-9 tracks the prices for three products from source to end markets; one may note a substantial mark-up, 40–80 per cent generally, at each new stop. Various field reports note that mark-ups of 100 per cent or more are common every time a product changes hands. For traditional products that are now prohibited, suppliers complain that buyers force them to accept lower prices. Traders in turn complain of increased risk and higher costs ('facilitation fees') involved in moving such contraband.

Merchants are nevertheless key elements in the trading system. Traders play an important role in collecting, grading, storing and shipping goods, though some would argue that their rewards are disproportionate to their contribution, either to commerce itself or to economic development in general. Throughout Southeast Asia it is not uncommon to find trade networks dominated by resident foreigners prospering by furnishing the financial and managerial skills needed to lubricate peasant economies lacking institutional arrangements required for market development. In Indochina, resident Chinese dominated the nucleus of local entrepreneurs at the beginning of the 20th century (Norlund 1991: 86). In Laos, the Chinese were joined by the Vietnamese, originally brought in by the French to run the colonial administration (Gunn 1990). Both Vietnamese and Chinese traders are still active in the export networks operating presently in southern Laos (Baird 1994; Enfield et al. 1998). Villagers in remote areas of Laos and also in other parts of Southeast Asia report that traders from China now come directly to the village to discuss with local farmers the availability of a variety of forest products. The social networks if the commercial community, regulated by custom and long-term relationships, demonstrate a persistence through time, including in periods of dramatic political upheaval and economic change, that is testimony to their relative political autonomy as well as internal strength (Barney 1967; Evers 1991; Hill 1998).

As discussed above, many forest products were regulated as state monopolies. The revenues from the trade in these products plus taxes on commerce in general provided a significant source of revenues for the ruling elite. This traditional linkage between resource exploitation and fiscal revenues persists in modern systems. Stumpage fees or royalties for log extractions are common throughout the region. However, they are often set at unrealistically low levels and, as a result, as in the case of Malaysia (Leigh 2000) or Indonesia (Abe, this volume), the primary beneficiaries are often the family members or the political cronies of elites. Military involvement has also been implicated in logging operations throughout the region. Such is the case in Laos (Stuart-Fox 1995). According to the World Bank (1997) unrecorded logging by firms associated with the military and private road construction are two of the biggest threats to forest resources in Laos.

Finally, it is a mistake to limit the analysis to the activity of Lao citizens. International boundaries are no barriers to either resource exploitation or trade. Collectors from neighboring countries regularly cross into Laos to harvest non-timber forest products; merchants in Vietnam argue that most

Table 3-9. Prices in various markets for four products originating in Laos, ca. 1898

MARKET	PRICE (IN FRANCS PER 100 KG)			
	STICK LAC	BENZOIN	CARDAMOM	RUBBER*
Hua-Phon (Laos)	n.a.	n.a.	n.a.	1st 166–208 2nd 85–115
Nong-Khay, Vientiane (Laos)	50–60	150–250	n.a.	1st 250–280
Bassac (Laos)	38–40	n.a.	92–100	n.a.
Korat (Thailand)	75–95	250–280	85–110	n.a.
Phnom-Penh (Cambodia)	54–70	n.a.	133–165	1st 540–625
Bangkok (Thailand)	125–135	300–350	140–180	n.a.
London (U.K.)	220+	500–700	250–310	1st 750–835
Hamburg (Germany)	n.a.	n.a.	n.a.	1st 800–1000 2nd 450–600

Source: Derived from Reinach, 1901, p. 429

Note: n.a. = not available; * = two grades listed

of the wild animals that they sell are not 'illegal' because they do not come from Vietnam, but rather Laos (Le ca. 1995, Timmins 1995, MacDonald 2001). Foreign investments in the forest sector contribute to degradation as opposed to development in that they generally concentrate on activities that facilitate extraction, e.g., road-building. Thai, Chinese, Vietnamese, Taiwanese and Malaysian timber companies have benefited substantially from logging concessions in Laos.

Conclusions

It is now generally recognized that forests serve a broad spectrum of society's needs, providing basic necessities for many inhabitants as well as foreign exchange for the government treasury. Forest products, a unique reflection of the natural biota in an area, have been the basis of exchange relationships between neighboring societies since earliest times. Important as items of tribute and, later, trade, forest products have provided the assets that various groups needed to strengthen their position, both economically and politically. In examining closely the export rosters of the ancient trading capitals of Southeast Asia we have seen that many of the products for which they became famous may well have come from Laos. These products, both consumer goods and raw materials, served to fuel the development of these societies that received them as well as the trading societies through which they transited. At the source collectors and local producers, often pressed into supplying these goods by physical threat or a financially punitive tax system, by contrast seem to have benefited considerably less. From architectural remains and historical artifacts one could conclude that the ruling elite in these societies invested in religious as opposed to social development.

Recognizing the data limitations and the preliminary nature of this analysis, this work has focused on comparing historical accounts with observable current conditions (as observed in field studies) within Laos. With regard to forest products exploitation and trade, it seems that many aspects have changed very little over time. Most patterns of trade have persisted in substantially the same form, though product composition has been enhanced as a result of technological change. With additional products in the mix we see a stratification of trade with many traditional products still traded, but in the informal sector. The direction of forest products trade has remained largely the same, from the have-nots to the have, raw materials exchanged for manufactured goods.

With regard to distribution of costs and benefits (both economic and environmental), historical patterns appear to persist. From all appearances it is largely the intermediaries that grow rich. Laos remains one of the poorest (economically) and least developed nations in Southeast Asia. The government has been unable or unwilling to enforce environmental protection regulations and as a result the natural riches are being appropriated and sold off piecemeal for largely private benefit. The most notable change has been in the pace of trade, which has had significant impact on environmental conditions. As the tempo of trade intensifies with the application of new technology dramatically, the pressure on resources increases. Once common species, such as the wild elephant and the tiger, have become rare and are on the verge of extinction.

Increasingly linked to global markets, the trade in forest products appears to be both demand and supply driven. In other words, the increased demand for many products as a result of an expanded market is complemented by the suppliers' increased demand for cash. Ultimately, as Laos gets absorbed into the global economy it appears that the fate of the forest rests not only in the hands of the country itself but also on what develops outside. External demands coalesce with internal pressures to create a pattern of exploitation that often seems oblivious to natural laws

(Abe, this volume). Consumerism has been so successfully promoted that the raw materials are being exhausted, yet societies resist institutional strictures to constrain exploitation.

Finally, we must conclude that the history of forest products exploitation and trade in Laos has been and unfortunately continues to be essentially a mining operation. That is, extraction is the focus and, except for a few species, such as benzoin or bamboo, little investment has been made in developing sustainable production systems for the products or the people who have contributed substantially to the long-term development of this region. Moreover, little of the proceeds of forest exploitation appear to have been invested in developing alternatives, which might have significantly reduced economic dependence on natural resource extraction. The movement of goods from the uplands to the lowlands reflects not only a shift in resources, but a shift in power as the forest and its potential is depleted. But this experience is not limited to Laos; it is common across the region. Indeed it reflects the relationship between countries in region as a whole, between Laos and its neighbors. Without more concerted attention to developing socially equitable and environmentally sound production systems for the wide variety of forest products, upland peoples will find not only their economic livelihood at risk but their culture threatened. Lacking such initiatives, it may well be true that Laos will be remembered not only for the great variety of forest plants and animals traded there, but for 'trading in' her forests for very little benefit.

Notes

1 Lao People's Democratic Republic is the official name of the country but government officials inform me that it is also acceptable to use the shorter name, Laos.
2 Although Marco Polo reportedly entered the northern reaches of Lao territory, his publications are full of inaccuracies and many claim they are based on hearsay rather than direct observation (Hall 1968:228).
3 Average wages in Lao PDR although 32 per cent higher than those in China and 44 per cent higher than those in Vietnam are less than half those in Thailand (FEER 1998). These figures hardly reflect the true situation as the vast majority of the people do not enter the labor market.
4 By the 19th century southern China had so developed its export specialties, including silk, sugar, tea and tobacco, that the landscape had been virtually

stripped of its tropical forests pushing out many of the wild species, such as the tiger, that had once lived there (Marks 1998).
5 Specifically, the English in India, Burma, Malaya and north Borneo; the Dutch in the East Indies, the French in Indochina, the Spanish in the Philippines (until 1898 when the Americans took over) with the Portuguese in the end limited to a few isolated islands and coastal trading posts.
6 Apparently these animals were still quite common in Indochina at the end of the 19th century as reports from numerous travelers testify, e.g., Mouhot 1868.
7 Other east-west routes connecting Laos to Vietnam are currently (2001) under renovation.

Bibliography

Baird, Ian. 1994. The Trade in Soft-shelled Turtles (Trionychidae) between Southern Lao PDR and Vietnam. Unpublished report. TRAFFIC (Southeast Asia), Kuala Lumpur.

Barden, A., N. Awang Anak, T. Mulliken, and M. Song. 2000. Heart of the Matter: Agarwood Use and Trade and CITES Implementation for *Aquilaria malaccensis*. Cambridge: TRAFFIC International.

Barney, G. Linwood. 1967. The Meo of Xieng Khouang Province, Laos. In Peter Kunstadter, ed., *Southeast Asian Tribes, Minorities and Nations*, 271–294. Princeton: Princeton University Press.

Berval, Rene de. 1959. *Kingdom of Laos*. Saigon: France-Asie.

Blaikie, Piers. 1985. *The Political Economy of Soil Erosion in Developing Countries*. London: Longman.

Breazeale, Kennon, ed. 1999. *From Japan to Arabia: Ayutthaya's Maritime Relations with Asia*. Bangkok: The Foundation for the Promotion of Social Sciences and Humanities Textbook Project, Toyota Thailand Foundation.

Bureau de la Statistique Generale. 1937. *Annuaire Statistique de l'Indochine, 1934–1937* 6. Hanoi : Direction des Affaires Economiques et Administratives, Gouvernement General de l'Indochine.

Bryant, Raymond L. and Sinead Bailey. 1997. *Third World Political Ecology*. London: Routledge.

Bulbeck, David, Anthony Reid, L.C. Tan, and Y. Wu. 1998. *Southeast Asian Exports Since the 14th Century: Cloves, Pepper, Coffee and Sugar*. Singapore: Institute of Southeast Asian Studies.

Carne, Louis. 1872. *Travels on the Mekong, Cambodia, Laos and Yunnan.* Bangkok: White Lotus. Reprinted 1995.

Chang, Pin-tsun. 1991. The first Chinese Diaspora in Southeast Asia in the 15th Century. In Roderich Ptak and Dietmar Rothermund, eds., *Emporia, Commodities and Entrepreneurs in Asian Maritime Trade, c. 1400–1750*, 13–28. Stuttgart: Steiner Verlag.

Chang, Stephen Tseng-Hsin. 1991. Commodites Imported to the Chang-chou Region of Fukien during Late Ming Period: A Preliminary Analysis of Tax Lists Found in Tung-hsi-yang k'ao. In Roderich Ptak and Dietmar Rothermund, eds., *Emporia, Commodities and Entrepreneurs in Asian Maritime Trade, c. 1400–1750*, 159–194. Stuttgart: Steiner Verlag.

Coedes, G. 1967. *The Making of Southeast Asia.* Berkeley: University of California Press.

Crawfurd, John. 1830. *Journal of an Embassy from the Governor-General of India to the Courts of Siam and Cochen China* 2. 2nd Edition. London: Henry Colburn & Richard Bentley.

Cushman, Jennifer W. 1993. *Fields from the Sea : Chinese Junk Trade with Siam during the Late Eighteenth and Early Nineteenth Centuries.* Studies on Southeast Asia 12. Ithaca: Southeast Asia Program, Cornell University.

Donovan, Deanna G. 1998. *Policy Issues of Transboundary Trade in Forest Products in Northern Vietnam, Lao PDR and Yunnan, PRC.* Proceedings of a workshop, September 14–20, 1997, Hanoi, Vietnam. Washington, D.C.: World Resources Institute.

––––––– 1999. *Strapped for Cash: Asians Plunder their Forests and Endanger their Future.* Asia Pacific Issues Series 39. East-West Center, Honolulu, Hawaii.

Elson, R.E. 1997. *The End of the Peasantry in Southeast Asia: A Social and Economic History of Peasant Livelihood, 1800–1990s.* New York: St. Martin's Press.

Enfield, N.J., Bandith Ramangkoun, and Vongvilay Vongkhomsao . 1998. *Case Study on Constraints in Marketing of Non-timber Forest Products in Champasak Province, Lao PDR.* NTFP Project, Department of Forestry, Vientiane.

Evers, Hans-Dieter. 1991. Traditional Trading Networks of Southeast Asia. In Karl R. Haellquist, ed., *Asian Trade Routes.* Copenhagen: Scandinavian Institute of Asian Studies.

Fillastre, Adrien. 1905. Bois d'Aigle et Bois d'Aloes. *Revue Indo-chinois* (1st sem.), 248–262.

Foppes, Joost and Sounthone Ketphanh. 2000. Forest Extraction or Cultivation? Local Solutions from Lao PDR. Paper presented at: Workshop on the Evolution and Sustainability of Intermediate Systems of Forest Management, June 28–July 1, 2000, Lofoten, Norway.

Foppes, Joost, Thongphoune Saypaseuth, Khamsamay Sengkeo, and Seng Chanthilat. 1997. *The Use of Non-timber Forest Products on the Nakai Plateau, Khammouane Province Lao PDR*. NTEC (April 1997). Vientiane, Lao.

Gunn, Geoffrey C. 1990. *Rebellion in Laos: Peasant and Politics in a Colonial Backwater*. Boulder: Westview Press.

─────── 1999. *Nagasaki in the Asian Bullion Trade Networks*. Re. Bull. 32. Research Instittue of Southeast Asia, Nagasaki University.

Hill, Ann Maxwell. 1998. *Merchants and Migrants: Ethnicity and Trade among Yunnanese Chinese in Southeast Asia*. New Haven: Yale University Southeast Asia Studies.

Hirth, F., and W.W. Rockhill. 1967. *Chau Ju-Kua: His on the Chinese and Arab Trade in the Twelfth and Thirteenth Centuries, Entitled Chu-fan-chi*. Taipei: Ch'eng-wen Pub. Co.

Ho, Ping-yin. 1935. *The Foreign Trade of China*. Shanghai: Commercial Press Ltd.

Hu, Shiu Ying. 1990. History of the Introduction of Exotic Elements into Traditional Chinese Medicine. *Journal of the Arnold Arboretum* 71:487–526.

King, Victor T. 1995. *Explorers of South-east Asia*. Kuala Lumpur: Oxford University Press.

Latham, A.J.H. 1997. Modernisation and Progress :Southeast Asia in the World Economy, 1860–1941. In *Proceedings of the International Symposium on Southeast Asia : Global Areas Studies for the 21st century. October 18–22, 1996*, 207–224. Center for Southeast Asian Studies, Kyoto University.

Laurent, Chazee. 1990. *La Province d'Oudomsay*. Vientiane: UNDP.

Le Dien Duc ca. 1995. *Information about Agarwood Trade in Vietnam*. CRES, University of Vietnam, Hanoi.

Lebar, F.M. and A. Suddard. 1960. *Laos, Its People, Society and Culture*. New Haven: HRAF Press.

Li, H. 1979. Translation and Commentaries. In *Nan-Fang Ts'ao-Mu Chuang: A Fourth Century Flora of Southeast Asia*, 87–91. The Chinese University Press. Hong Kong.

Li, Tana. 1998. *Nguyen Cochinchina: Southern Vietnam in the Seventeenth and Eighteenth Centuries*. Southeast Asia Program, Cornell University, Ithaca, New York.

Ma Huan. 1433. *Ying-yai Sheng-lan: The Overall Survey of the Ocean's Shores*. Translated with introduction, notes and appendices by J.V.G. Mills. Reprint 1997. Bangkok: White Lotus Co. Ltd.

MacDonald, Phil. 2001. Laos Undercover. *Asiaweek* 31 August 2001.

Manarungsan, Jomsop. 1989. *Economic Development of Thailand, 1850–1950*. IAS Monograph 042. Bangkok: Institute of Asian Studies, Chulalongkorn University.

Manguin, Pierre-Yves. 1972. *Les Portugais sur les Cotes du Vietnam et du Campa: Etudes sur les Routes Maritimes et les Relations Commerciales, D'apres les Sources Portugaises (XVIe, Xviie, XVIIIe siecles)*. Paris : Ecole Francaises d'Extreme Orient.

Marini, G. F. de. 1663. *A New and Interesting Description of the Lao Kingdom*. Translated by Walter E.J. Tips (1998). Bangkok: White Lotus.

Marks, Robert B. 1998. *Tigers, Rice, Silk and Silt: Environment and Economy in Late Imperial South China*. New York: Cambridge University Press.

Ministry of Finance, Laos. 2000. *National Statistics, 1999*. Vientiane.

Momoki, Shiro. 1998. Was Champa a Pure Maritime Polity? Paper. Seminar on Eco-history and the Rise/Demise of the Dry Areas in Southeast Asia, October 13–16, Center for Southeast Asian Studies, Kyoto, Japan.

Mouhot, Henri. 1868. Ferdinand de Lanaye, ed., *Voyage dans les royaumes de Siam, de Cambodge, de Laos et autres parties centrale de l'Indo-Chine*. Paris: Hachette.

Nash, Stephen, ed. 1997. *Fin, Feather, Scale and Skin: Observations on the Wildlife Trade in Lao PDR and Vietnam*. TRAFFIC-Southeast Asia, Kuala Lumpur.

Nash, Stephen and S. Broad. 1993. Guidance on the assession of Lao PDR to CITES and other conservation treaties. IUCN, Vientiane.

Noone, R.O.D. 1954. Notes on the Trade in Blowpipes and Blowpipe Bamboo in north Malaya. *Federation Museums Journal* I–II: 1–18.

Norlund, Irene. 1991. The French Empire, the Colonial State in Vietnam and Economic Policy, 1885–1940. In G.D. Snooks, A.J.S. Reid and J.J. Pincus, eds., *Exploring Southeast Asia's Economic Past* [*Australian Economic History Review* XXXI(1):72–89].

Padoch, Christine. 1992. Marketing of Non-timber Forest Products in Western Amazonia: General Observations and Research Priorities. In Daniel Nepstad and Stephan Schwartzman, eds., *Non-timber Forest*

Products from Tropical Forests, 43–50. *Advances in Economic Botany* 9. Bronx: New York Botanical Garden.
Polanyi, Karl. 1975. Traders and Trade. In *Ancient Civilisations and Trade* edited by J.A. Sabloff and C.C. Lamberg-Karlovsky. Albuquerque: University of New Mexico Press.
Provencher, Ronald. 1975. *Mainland Southeast Asia: An Anthropological Perspective*. Pacific Palisades, California: Goodyear.
Purcell, Victor. 1951. *The Chinese in Southeast Asia*, Second Edition. Kuala Lumpur: Oxford University Press. 1991 reprint.
Reid, Anthony. 1993. *Southeast Asia in the Age of Commerce, 1450–1680. Vol. II: Expansion and Crisis*. New Haven: Yale University Press.
Reinach, Lucien de. 1901. *Le Laos* 1. Paris: A. Charles Libraire-Editeur.
Salter, Richard E. 1993. *Wildlife in Lao PDR: A Status Report*. Vientiane: IUCN.
Schaller, George. 1998. On the Trail of New Species. *International Wildlife* July/August: 37–43.
Service de la Statistique Generale. 1927–1937. *Annuaire Statistique de l'Indochine* 1. Recueil de Statistiques, 1913–1922. Hanoi : Direction des Affaires Economiques, Gouvernement General de l'Indochine.
Skinner, G. William. 1964. Marketing and Social Structure in Rural China. *Journal of Asian Studies*. AAS Reprint Series 1.
Stuart-Fox, Martin. 1995. *Southeast Asian Affairs, 1995*. Singapore: Institute of Southeast Asian Studies.
———— 1997. *The History of Laos*. Cambridge: Cambridge University Press.
Suksavang Simana. 1997. *Kmhmu' livelihood: Farming the Forest*. Translated by Elizabeth Preisig. Vientiane: Institute of Cultural Research, Ministry of Information and Culture, Laos.
Timmins, Robert J. 1995. Concerns of Wildlife Trade in Lao PDR. Paper presented at Workshop on Wildlife Trade in Lao PDR. June 21–22, 1995, Vientiane. CPAWN, Department of Forestry, Vientiane.
Viraphol, Sarasin. 1977. *Tribute and Profit: Sino-Siamese Trade 1652–1853*. Cambridge: Harvard University Press.
Wang, Gungwu . 1958. The Nanhai Trade: A Study of the Early History of Chinese Trade in the South China Sea. *Journal of the Malayan Branch of the Royal Asiatic Society* 31:1–135.
Wheatley, Paul. 1959. Geographical Notes on Some Commodities Involved in Sung Maritime trade. *Journal of the Malayan Branch, Royal Asiatic Society* 32(2): 5–140.
———— 1979. Urban Genesis in Mainland Southeast Asia. In R.B. Smith

and W. Watson, eds., *Early South East Asia,* 288–303. New York: Oxford University Press.

White, John. 1823. *History of a Voyage to the China Sea.* Boston: Wells and Lilly.

Wicks, Robert S. 1992. *Money, Markets and Trade in Early Southeast Asia.* Ithaca: Southeast Asia Program, Cornell University.

Winichakul, Thongchai. 1994. *Siam Mapped.* Honolulu: University of Hawaii Press.

Wolf, Eric R. 1982. *Europe and the People Without History.* Berkeley: University of California Press.

Wolters, O.W. 1982. *History, Culture and Religion in Southeast Asian Perspectives.* Singapore: Institute of Southeast Asian Studies.

Woodside, Alexander. 1995. Central Vietnam's Trading World in the Eighteenth Century as seen in Le Quy Don's 'Frontier Chronicles.' In K.W. Taylor and K. Whitmore, eds., *Essays into Vietnamese Pasts,* 157–177. Studies on Southeast Asia 19. Ithaca: Southeast Asia Program, Cornell University.

World Bank. 1997. Lao PDR Public Expenditure Review. Report no. 16094-LA. East Asia and the Pacific Regional Office, Bangkok.

4
The Political Ecology of Forest Products in Indonesia: A History of Changing Adversaries

Wil De Jong, Brian Belcher, Dede Rohadi,
Rita Mustikasari, Patrice Levang

Forest products (FP[1]) have been traded internationally from Asia's tropical forests since somewhere between the third (Peluso 1983) and fifth centuries (Andaya and Andaya 1981). Over the past decades, development, conservation, and indigenous rights advocates have shown interest in FP for different reasons. Until the mid-1980s development specialists expected that *minor forest* products could contribute to improve national incomes, or incomes among the rural poor living in or near forests. Since the late '80s, conservation advocates perceived collecting of *non-timber forest products* as a possible alternative to destructive logging practices. Indigenous rights activists have argued for the return of forest control to the resident communities that traditionally collected products for consumption and sale.

Several researchers have questioned whether it is appropriate to commercialise FP for development or conservation. Some of the arguments advanced against further commercialisation include the use of coercive measures by some entrepreneurs to force collectors to extract products with wide international demand, as happened during the rubber boom around the turn of the century (see Hardenburg 1912; Weinstein 1983). Others (for example, Dove 1993, 1996) argue that successful commercialisation will necessarily lead to the larger share of the profits being siphoned off by well-connected entrepreneurs, who will then be in an even stronger position to take over control of the trade. Thus, the primary forest product producer will eventually benefit less and less.[2] This process necessarily leads to stagnation or a decline in income among the primary FP producers. In addition, it often leads to unsustainable use of natural resources (Neuman 1996). Hence, the proposition that increased commercialisation of FP will decrease deforestation for alternative land use is problematic (see de Jong 1999). When conditions of coercion, or resource appropriation by outsiders occur, the producers have no incentive left to maintain forest that yield commercially viable FP while outsiders are usually inclined to exhaust resources and have no concern for long-term supply.

FP are, by nature, particularly prone to market failure. They are typically produced in relatively small quantities, in remote areas, and are often "open access" resources—the resource managers have little control over the resource. FPs often have sharp seasonal fluctuations owing to the phenology of plants, migration patterns of animals and climatic conditions. Thus, transactions cost are high and the bargaining power of producers is low (Belcher 1998).

This paper pursues a political ecology[3] analysis of the history of contestation over procurement and benefit-capture of FP in Indonesia. This analysis of historical events falls within what Bryant (1992) distinguishes as one of the "critical areas of inquiry" of political ecology. Political ecology rightly introduces variables like *power relations* and the *political influence of economic elites* to explain actions of FP procurement and trade, thus avoiding simplified economic reductionism (Bryant 1992). By critically analyzing the nature of the actions that participants in FP economies undertake to increase their share of profits, political ecology inquiry adds theoretical depth to criticisms of the alleged development and conservation potential of FP commercialization (see, for example, Arnold and Ruiz Perez 1998). This analysis also has practical implications for the kind of political-institutional actions that need to be undertaken when FP-based income improvement is attempted.

This paper will look at FP procurement and trade in Indonesia in a historical perspective, and critically investigate the nature of the actions of the players involved in these economic activities. Section two introduces our analytical approach to information found in the literature. Sections three and four provide a brief overview of the main historical periods of Indonesia and how FP procurement and trade changed during these periods. We will then go through four examples of commercially exploited FP that were prominent during much of this economic history. The last section of the paper will discuss the findings in the light of the categories of rent-seeking proposed in section two.

The Analytical Approach Followed in This Paper

Central to this paper is a classification of the actions of the players in FP procurement and trade to increase the benefits they derive from their participation in this business. The central concept in the analysis below is rent seeking. Rent is understood here as a surplus profit beyond the normal market return to labor and capital, which may be a result of scarcity of a resource, regulations, or monopoly. Rent seeking may include the creation of rent, or the capturing of rent. In the perfectly competitive economy, there would be no rent, except for a natural scarcity rent. This is a situation that is a hypothetical ideal rather than a reality in life. Many of the actors in a market economy and their political allies will engage in putting up regulations that allow them to create or capture rent. It is part of the political game and it reflects the struggle between different interest groups, which often is defined by the difference in power that the groups have. However, this definition provides and adequate criteria to distinguish the kind of actions that we want to identify throughout the history of FP procurement and trade in Indonesia. Using this definition, common rent seeking strategies are, for instance collusion between a small group of traders, withholding information, coercion in the sense of using or threatening with physical force, and the like.

As this effort is largely of an inventory kind of exercise, this paper does not pretend to develop a perfect theoretical coherent classification of rent-seeking actions, as per the definition above. Using the criteria resulting from the general definition presented in the previous paragraph, this paper reviews important examples in the history FP procurement and trade in Indonesia, identifies the different actions of the various participants that can be qualified as rent-seeking, and proposes a classification of these actions in different categories. In the case of procurement and trade of FP from Indonesia the paper distinguish the type of actions listed below:
1. Those with power engage in coercion through threatening with physical violence or imprisonment.
2. Traders in the regions where forest products are collected engage in cooperation and collusion.
3. Those close to central government influence policies or legislation.
4. Provincial and district officials benefit from inadequate enforcement of rule of law from central government.
5. Actors benefit from ill defined property rights.

Using these different types of rent seeking allows the identification of changes in strategies used by different actors in order to increase their benefits from FP trade, and some changes throughout history of rent seeking strategies

Of Kings, Sultans, and Forest Product Protectors

The Benign Role of the Usual Villains

Trade in FP from Indonesia started as early as the third century AD. (Peluso 1983). As products had to be transported by sea, Chinese and Bugis merchants controlled trade networks for much of the pre-colonial and colonial era in eastern Southeast Asia, whereas several Malay kingdoms controlled the trade towards the western part of the region (Andaya and Andaya 1981). The nature of the ownership regimes that could have limited primary collectors' share of the benefits varied according to the product, the place, and the era in which they were collected. Sandalwood from the eastern islands, for instance, has been traded to China since the 10th century, and became of interest to European buyers since the 15th century (Rohadi et al. 2000). During this era, local rulers controlled sandalwood in Timor and possibly in the surrounding islands. Traditionally, tribal rights of property over the resource were recognized, but the king claimed his share of the profits. Different regions had their own kings and each claimed sovereign ownership of sandalwood resources. These kings would appoint a "landlord" to control sandalwood in their particular regions. Landlords in turn appointed a village leader in each village whose duties included the control of sandalwood. The king had rights to the sandalwood roots, the landlord to the stem, and the landowner, the individual producer, to the branches (Ormeling 1955). If intruders collected sandalwood outside their own territories, this could initiate a war between tribes and usually ended up with a severe fine levied on perpetrators (Tapatab 2000).

This kind of control of forest product exploitation was not universally common in the archipelago. The coastal economy of Borneo, which for centuries was mainly based on smelted iron and gold, and on forest product extraction and trade (Andaya and Andaya 1981; Peluso 1983), was dominated by sultanates and what Sellato (2000) identifies as "petty" kingdoms. Similar territorial and resource control organization was found in West Borneo and Sumatra. Vijayapura in West Borneo was a major port that existed together with the well-known Srivijaya kingdom, which flourished in western Sumatra between the 8th and 12th centuries (Andaya and Andaya 1981). Sultans and kings from the eastern parts of Borneo had established themselves in the region from the 13th century onwards and mostly came from Java (Peluso 1983). They settled at crucial locations along major rivers or estuaries and largely engaged in taxing those who shipped products from the forests to faraway locations. This coincided with a change in the regional trade politics that was largely dominated by China's recognition of Srivijaya as its main Southeast Asian trading partner. The new policy within China encouraged establishment of many new ports in the archipelago and also coincided with the establishment of P'o-ni, supposedly the forerunner of Brunei port in northwest Borneo (Andaya and Andaya 1981).

The sultanates in Borneo built an important part of their wealth on trade and on duties levied on trade goods, by taking full advantage of their geographical positions along the transportation channels between upstream collection areas and the final destinations (Sellato 2000). The sultanate of Bulungan, East Kalimantan, for instance, had hardly any access to territory. However, it managed to force tribal groups near its vicinity to submit to collecting and farming for the sultan. On the other hand, upstream groups remained autonomous and eventually rebelled when the sultan tried to force them to direct their trade his way, impose heavy duties, or pay tribute in the form of rice and FP. The ruling Kutai, another sultanate in East Kalimantan, levied taxes on any products that went through its territories.

Trade in FP from East Kalimantan intensified from the 17th century onwards. This coincided with an overtaking of the control of the trade of the northern half of Borneo by the Sulu kingdom, formerly a vassal of Brunei (Andaya and Andaya 1981). The trade was conducted by the Chinese and several ethnic groups from the archipelago, but was later dominated for much of the history by the Bugis, originally from Sulawesi. The Bugis initially had spread to Java, Palembang and Jambi, after conflicts in Sulawesi partly fostered by the Dutch to increase their control of the island and the region (Andaya and Andaya 1981). These refugees came into conflict with local authorities and eventually moved on to remoter places like the southwest and east coast of Borneo. The Bugis traders gained considerable political power because of their seafaring skills and trade connections. Eventually the Dutch were to take control of the FP trade, but not until the 20th century.

There is little evidence of any coercive methods to engage people living farther inland in the harvesting of products. Coastal rulers or traders and inland collectors did not dispute property rights over the resources. Partly, interior groups managed until recently to keep outsiders away from their territories, although they did fight among themselves over the resources that were in demand downstream (Sellato 2000). The Kutai, for instance, operated for a long time as middlemen between the interior Dayak and the Bugis traders. They apparently went upriver to trade themselves, sent their agents, or received goods sent by the inland chiefs. The sultans levied taxes, not so much because they considered themselves owners of the FP, but rather because they controlled vital parts of the transportation system.

The interior groups became heavily involved in FP exploitation and trade after 1850, and the mostly remote groups in this region since the last decades of 19th century, when demand for certain products (like *gutta percha*) intensified (Sellato 2000). While this happened, trading posts moved upstream to tap ever more remote regions. The petty kingdoms, especially, moved inland as they opened remote areas to trade and bring into partnerships with them leaders of more isolated tribal groups. To get access to trade resources, these kingdoms would strike deals with tribal chiefs rather than force them into trade. For the petty kingdoms often competed among themselves to set up favorable partnerships with those inland chieftains that held exclusive access to resource-rich territories.

We have little evidence how these deals were sealed, if contracts were used or not, if agreements were breached, and how parties to agreements made efforts to minimize risks and losses. Marriage and blood brotherhood were common ways to bind tribal groups to downriver kingdoms (Sellato 2000). Not all these efforts were successful, partly because of interethnic struggles over control of forest resources in the interior. The Taogus, another coastal group of the region, for instance, also tried to intermarry with the native women to increase control of birds-nests caves. This was largely unsuccessful because, by that time, the Kenyah, one of the more powerful interior Dayak groups, started to control some of the forest product resource areas through coercive measures. As a result some tribes who had collected for the Taosug became too scared of the Kenyah to continue going into the forest (Peluso 1983). Efforts to use coercive measures by downstream parties reportedly were largely unsuccessful. The Taosug, in addition to finding alliances, also attempted to control "host" populations, in the interiors where they settled to control FP trade. They used raiding for the sake of trade but eventually were themselves "raided out" by the Kenyah (Peluso 1983). Taosug are famous today for one of highest rates of feuding and homicide!

On the other hand, several of the groups interested in FP made use of slaves for their collection. Dutch accounts from the 17th century report the use of slaves among Kenyah, Taosug and Bugis to collect FP in East Kalimantan (Peluso 1983). The petty-kingdoms even provided tribal chieftains with slave labor procured directly through raids or indirectly from the maritime slave trade, in order to intensify FP procurement. The formal abolition of slavery dealt a final blow to the kingdoms' economy and to the tribal chieftains, for whom slaves were a major asset. Some, however, continued to make use of the Punan hunter-gatherers who lived in FP collection territories. Those patrons defended their Punan by diplomacy and by force. However, eventually Punan also started trading directly with downstream traders (Sellato 2000).

Bugis, Chinese, or Arabs started settling in remote upstream villages since 1920s, 1930s, and even 1950s. The Bugis did manage to establish direct relationships with the inland Dayak chiefs, a condition that allowed them to negotiate on their own terms. As the Bugis controlled the import of "foreign" goods, they were able to pressure for lower prices by delaying delivery of salt, an indispensable commodity for the inland population (Peluso 1983).

Visitors Turning into Desperados

The Dutch followed some general principles when they established the rule of law in the vast territory that is now Indonesia. The Dutch claimed to govern the "natives" of Indonesia based on customary law or *adat*. According to Vargas (1985), this was an improvement, at least in Java, over previous governance systems as the old kings had paid little attention to customary law. On the other hand, the Dutch divided the territories into governed and self-governed lands (Vargas 1985). Most of the remote FP collecting regions belonged to the self-governed lands where governance basically was not changed from as the system that was already in place before the Dutch took over. The need to recognize customary law or to replace it with a western legal system became a hotly debated issue. The customary law of Borneo, however, never reached the same recognition as in Java and Sumatra (Vargas 1985).

It was not until the 1930s that the Dutch achieved a "semblance of control" in the upstream regions of East Kalimantan (Geary and Eaton 1992). Once they did so, the Dutch made detailed lists of bird nests caves and their owners, thereby formally recognizing de facto ownership. Some of the coastal rulers obtained recognized ownership of such caves, which they continue to hold today. During the period 1880–1920 traditional trading channels continued side by side with the emerging Dutch-controlled trading channels, as the Dutch were unable to impose their rule of law and claim taxes. They did not have the resources to back up these claims. In the 1920s, however, the bulk of the FP trade came under Dutch administration (Sellato 2000).

In eastern Indonesia, colonial visitors became interested in sandalwood resource much earlier. As stated above, the Portuguese and Dutch were involved in the trade from the 15th century onwards (Husain 1983). At the beginning, the traders were using the kings or tribal heads to obtain the wood. It seems that during the beginning of the colonial era, there was competition between the Portuguese, Dutch, Chinese, Javanese, Bugis, and Malay traders, but the local rulers continued to control the extraction.

After they had firmly established themselves in Indonesia, the Dutch colonial government monopolized the trading of sandalwood. Traditional rules were eliminated and replaced by the new colonial rules. Sandalwood extraction and trade was strictly controlled and monitored. The kings and head of tribes were still playing some roles in monitoring the resource but now they were themselves under the control of the colonial government. Severe financial or physical punishment was applied to any person who damaged a sandalwood tree without permission (Tapatab 2000).

A Lack of Solidarity

Ownership of natural resources is contested at many different levels, and between many different actors, including at the communal level. Opportunities to generate income from FP for local collectors are, in general, strongly influenced by the communal arrangements of access. Within Indonesia there are markedly different property regimes related to natural resources. Most of the marketed FP only can be held privately when, for instance, the relevant species are planted or are part of a managed forest garden. In some cases (as with, for example, honey trees in Kalimantan) ownership of forest trees can be established (see de Jong 2000).

A common feature is some kind of group property rights over tree or forest resources. In Kalimantan, for instance, most villages will claim ownership of a territory that often overlaps with a watershed. As a rule the resources in that territory are for the use of the village inhabitants while resource-rich lands in the territories often are strictly for village use only (Sellato 2000). The right of outsiders to harvest resources varied from place to place. A common rule in many locations is that outsiders may harvest resources for proper consumption but not for trade. The degree of exclusivity may vary per resource. In addition, ownership of resources among kinship groups is also common. These are mostly resource that were privately owned by one ancestor, and which were successively inherited by all the descendants (Peluso and Padoch 1996).

There are a few examples where local rulers have been capable of appropriating a resource for their exclusive use. The case of sandalwood, mentioned previously, is one example. Several of the East Kalimantan groups, like the Kayan, Kenyah, or Merap, have stratified societies that include aristocrats, commoners, and slaves. The aristocrats may hold exclusive rights over parcels of forestland (generally identified as *tanah ulen*) containing valuable resources like ironwood or rattan (Sellato 2000).

One forest product for which exclusiveness developed in response to outside demand were birds nests harvested from caves in East Kalimantan. Often the leader or the elites within the stratified groups appropriated these resources. Aristocrats, for example, had primary rights to the resource and could demand a share of the harvest. Although it is now not a strictly enforced rule anymore, some Kenyah villagers still give 5 per cent of the rattan they collect to the high ranking aristocrats (Momberg *et al.* 2000). In some cases neighboring groups could collect FP only with the permission of the aristocrats. For instance, Punan inhabitants in some locations could collect rattan in the territory of Kenyah villages, but they had to pay 20 per cent of their harvest to aristocratic leaders. The exclusive right to make decisions over forest resources privileged the aristocrats' rules. They could decide when to organize collecting trips of FP for the purpose of trade, as well as the trading trips. It implied that the petty kingdoms downstream made deals with these chiefs and nobody else. These chiefs got official recognition by the trading kings, the sultan, and the Dutch administration, for instance in the case of bird nests caves.

One consequence of the commercialization of FP was a growth in tribal warfare from 1830 onwards. Due to increased demand from the Chinese, a resource over which conflicts grew were the sites of bird nest (Peluso 1983). The groups who could extend their control over bird nests caves were the warlike groups, like the Kenyah and Merap, mentioned earlier. The Segai Kenyah, for instance, displaced weaker groups located close to the best caves (Peluso 1983). Much of the contemporary geographic distribution of groups in Kalimantan can be explained as a result of fights over and claims to the control of forest product resources.

The actions of these dominant groups were often part of in-fighting among different coastal trading groups. Although most products were collected and traded under unwritten agreements recognizing territorial exclusivity among traders, competition was also fought sometimes through collector groups. The Bugis traders, for instance, encouraged Kenyah raiders because they wanted to regain control of bird nests supplies that they had lost to the Taosug in Bulungan. However, even where this happened, the Kenyah kept the locations of caves a secret to outside traders.

When Generals Invade the Forests

The Ousting of the Kings

Indonesia declared independence in 1945. The first years after independence featured an ideology of reinforcing the rights and opportunities of those who had hitherto been oppressed. The new nation formulated a number of important laws and regulations to deal with its forests and forest resources it contained. Several of these laws created opportunities for legally recognized exploitation of FP by local groups. The constitution formulated soon after independence declared all natural resources as the property of the state, which was to be used for the maximum prosperity of the people (Weinstock and Sunito1989). Provincial and district governments were given the right to grant permission to collect timber and non-timber products in the new nation's forests. The era of sultans and kings was definitely over. Their roles in the FP trade had effectively ended. The non-indigenous traders also were looked at with suspicion, and they had little opportunity to continue their profession through the first years of independent Indonesia.

This progressive trend started to change with the Basic Agrarian Law of 1960. This law allowed the possibility of overruling customary law considerations whenever national interests were at stake. The Basic Forestry Law of 1967 called for research and development of FP for the betterment of the people living within or near the forest (Weinstock and Sunito1989). Government decree 21 of 1974 defined the conditions under which logging of timber could take place. It did create the opportunity for FP concessions of up to 100 ha and for periods of 6 months, providing these concessions did not hinder logging activities. A subsequent government decree revised number 21 of 1974, and did away with the FP concessions (Sirait pers.com[4]). Only in 1995, after much lobbying from the Indonesian NGOs and other advocates, was there a new legal instrument that allowed communal concessions for collection of FP.

Subsequent practices reflect this ambiguity in forest-related legislation. Since 1970, there has been an unprecedented exploitation of Indonesia's natural resources. Part of the trend involves adoption of increasingly coercive measures meant to transfer FP benefits away from those who previously had managed to make a reasonable living out of these sources. This process was most severe in the 1980s when a widespread onslaught on Indonesia's forest was fully under way.

The forests in which sandalwood was harvested had become formal state forest lands. Along with the progressive and liberating policies, punishment of local peoples for shifting cultivation practices that damaged trees was eliminated (Tapatab, 2000). In 1953, the new government announced a new Sandalwood Rule (*Peraturan Tjendana*). This rule encouraged local people to sustain the sandalwood resource. For the first time owners of the agricultural fields where sandalwood trees grew received the highest premium when harvesting the wood. The new payment for the landowners amounted to Rp 0.40 per kg. of sandalwood, while the kings and landlords only got Rp 0.05 and Rp 0.075 per kg. respectively (Ormeling 1955). At the time the price for a kg. of sandalwood was Rp 2.75 to Rp 5.50.

During the *orde baru* era, sandalwood became the major source of income in East Nusa Tenggara. The rules related to sandalwood management and production changed several times. One constant remained, and this was that the state continued to control the resource. Incomes collected from sandalwood sale were used for regional development. Subsequent changes in regulation mainly concerned benefit sharing of the harvested wood. The Regional Government Regulation No. 16 of 1986 permitted landowners to receive 15 per cent of the wood's value but only if they could prove their ownership with a certificate. In 1996, this landowner's share was increased to 40 per cent (Rohadi *et al.* 2000).

These measures seemed excellent in terms of social justice and respecting original ownership claims. However, as these policy measures were being formulated, practice on the ground was quite different. During the same period, illegal cutting of sandalwood grew as a result of the establishment of sandalwood oil companies with total capacity for production far above what was actually sustainable. This has been a typical phenomenon throughout the *orde baru* period and has affected resources like timber and rattan (see below). Although the Governor established levels of sustainable production based on inventories conducted by the Regional Forestry Office, lack of control by the government, and principally lack of political will to enforce regulations, for many years opened the door for corrupt officers to permit or support illegal cutting. The result has been a decline of the resource stock and a collapse of the sandalwood industry in Timor.

The Meddling of Tycoons

A similar story can be told about rattan. Rattan has been widely collected in Indonesia since 1835 (Sellato 2000). Somewhat later in South Kalimantan rattan suppliers began planting rattan in gardens (Belcher *et al.* 2000). Rattan trade and cultivation gained increased importance after independence and over the subsequent decades prices went up. By the end of the 1970s rattan had become the main source of income for many villages in regions of East Kalimantan.

In East Kalimantan, rattan was the raw material for manufacturing mats (*lampit*). This product gained increased international demand from 1985 onwards, mainly because of the use of *lampit* in Japan (Abe, this volume). Exports grew from an annual average of less than 1 million m² to 4.9 million m² in 1985. In line with other efforts to increase benefit capture of natural resources at home, the government decided to ban the export of unprocessed rattan. As a result the domestic processing industry expanded substantially. An additional consequence, however, was a drop in the price of the raw material, and thus in the income of rattan producers. In effect, the ban on exports of raw materials became a subsidy for domestic processors by reducing the total demand for rattan—at the expense of collectors' incomes.

Photo 4-1. Rattan has long been used by local peoples in East Kalimantan. In recent years well-connected business elites tried to control the international trade, as a result of which local incomes from this product have declined. *Photo by Wil de Jong*

Worse, however, was still to come. Eventually, the supply of *lampit* started to exceed demand while the excessive production led to a decline in quality and a corresponding decrease in price per unit. At this point one of Indonesia's best-known timber tycoons, Bob Hassan, saw his opportunity. He established the *Asosiasi Industri Permebelan dan Kerajinan Indonesia* (Association for Indonesia Furniture and Handicraft Industry, ASMINDO). In 1989 ASMINDO approached the Industry & Trade Department with a proposal to prevent unhealthy competition among rattan mat exporters. The proposal contained a trade policy, including a quota system to manage the supply. In addition it suggested imposing export restrictions. Since that time, traders have had to get approval from the ASMINDO office in Jakarta to export *lampit*, on top of which they had to pay a fee. Many individual manufacturers have since reported that the quotas were assigned based on political connections and payments.

The effects were devastating. In East Kalimantan, the 1989 ban on export of semi-processed rattan caused a decline in production, dipping from 13,500 tons in 1988 to 3,203 tons the next year and, finally, to 1,549 tons in 1991 (Sellato 2000). Although the price of raw materials remained the same, the demand went down and in many locations people could not sell their rattan any longer. For the next three years after the 1989-imposed regulations, *lampit* exports dropped to between 1.6 and 1.9 million m^2. It went up to 3 million m^2 in 1991/92 but dropped again to 1 million m^2 from 1996/97 onwards. The unit price of *lampit* dropped from US$ 6.38 to US$ 1.22 between 1987 and 1990, but went back up again to US 8.39 in 1995 (Belcher *et al.* 2000).

ASMINDO blamed Japanese consumers for this declining demand. Indeed, very soon after the export of *lampit* from Indonesia started to drop, imports from China, of mats made from bamboo strips, went up. The bamboo mats from China matches the export levels of rattan mats from Indonesia. This does suggest that the attempt by ASMINDO to control the industry actually favored conditions for a competitor using a close substitute to take over the trade (Belcher *et al.* 2000).

In the Name of Territorial Integrity

The more stunning example of coercive action related to FP is the case of *gaharu* incense wood, again from East Kalimantan. The *gaharu* trade in East Kalimantan started relatively later than in other regions because the quality was lower than that found in other collecting regions (Momberg *et al.* 2000). However, when supply of the higher quality gaharu from other places ran out, traders turned their attention to this region. Together with rattan, demand and prices for *gaharu* increased from the mid-'80s onwards.

In 1984–1985 the company PT Saguaro, owned by the Indonesian Army's Special Forces, Kopassus, came to East Kalimantan. They ordered all the *gaharu* traders to sell their entire stock to the company and simply forbade them to buy anymore from collectors. All the collectors had to sell exclusively to the company. Over night this company took over the entire *gaharu* trade by force and intimidation. However, even though they used coercion, the company did not stay in business long, to some extent because of lack of collaboration from the population.

The company also established bases in villages, "conscripted" traders, and forced people to go on *gaharu* collecting trips. In that time, no other FP were collected. During the 1990s, army helicopters, supposedly assigned to patrol the vast border area between Kalimantan and the neighboring Malaysian states of Sarawak and Sabah, exploited their access to airspace and dropped teams of collectors into uninhabited forest areas to extract *gaharu* for dispatch to Samarinda.

Will Justice Be Done?

Decentralisation, Democracy, and What Will Come of It

In April 1998 Suharto was forced to step down from power, marking the beginning of yet another period of profound change for the country. The post-Suharto years were marked by the call for *reformasi* (reformation) and an end to "KKN" (*korupsi, kolusi, nepotisme*).[5] These calls have led to a new way of doing politics in many parts of the country. They have paved the way to public discussion of issues, open expression of disagreement on current affairs among wide sectors of the country, and the bringing to justice of many abusive officials and business people, like Bob Hassan, who had enriched themselves during the Suharto regime.

The new government undertook revisions of the various legal instruments that were widely perceived as allowing the exploitation and degradation of the national resources of the country. Already the constitution has been amended to give more attention to private, economic, and socio-cultural rights. It also proposes genuine political autonomy of indigenous communities. In future amendments, adjustments are to be made to better define the rights of the state in relation to natural resources. Alternatives being weighed are whether the state should continue to have the right to assign resource ownership, or whether state powers should be limited to regulating administration. The outcome of these discussions may have important implication for other sector-based legislation.

Several proposals are on the books for new legal instruments that favor FP use among primary producers. One proposed government decree basically goes back to the former practice of recognizing customary rights. This would allow for legally recognized communal forest management and communal rights to harvest timber and non-timber, even in lands that continue to be classified as state forest land. This would be more applicable to traditionally managed forests, such as rattan gardens in Kalimantan. In addition, the new proposals allow communal concessions, again for both timber and non-timber, and small concessions for other legally entitled parties. The latter could be for timber or non-timber, but not for both simultaneously (Sirait pers. com.).

Unfortunately, political reform has coincided with one of the biggest ever economic crises in Indonesia and regionally. This had particular bearing on FP collection and trade because some of the planned options to stimulate economic recovery are the production of estate crops in remote areas of the country. Oil palm and industrial timber production are favored candidates for entrepreneurial investments; other crops like rubber are also slated (Potter and Lee 1998; Casson 2000).

On top of these developments Indonesia is finalizing a profound decentralisation program; the roots of this process began some time before the fall of Suharto. This again will have profound influences on how forests and their products are dealt with. There is still little evidence on the final outcome of this new tension between on the one a democratization that includes giving back the forest to the people, and on the other the political and economic decentralisation process, which gives *more* decision-making powers to local officials. The initial developments are worrisome, as the account below suggests.

Trees Versus Trees (Mayer 1998)
Already before the collapse of the Indonesian economy, estate crop development and industrial timber developments were on the rise. Production of oil palm is representative of this trend. Potter and Lee (1998) and Casson (2000) have in detail assessed the tremendous interests behind the oil palm boom and how this boom has led to a grab for land in several of Indonesia's outer islands. For instance, a large part of the province of West Kalimantan has already been given over as concessions

to oil palm companies, for future plantation development; current land uses or small farmer claims to the land are disregarded in this process (Potter and Lee 1998). By 1998, an estimated 70,000 ha were planted to oil palm in East Kalimantan (with substantially larger areas in neighboring provinces). Nearly 4 million ha have been designated for conversion in East Kalimantan and applications had been approved for more than 450,000 ha to be released by 1999 (Casson 2000). Provincial and district officials are interested in oil palm over other possible economic development options because it promises quick economic returns for the region (Potter and Lee 1998).

The companies interested in plantations proceed in different ways. From West Kalimantan there have been reports of the acquisition of land from local swidden agriculturists. This somehow suggests a de facto recognition of property rights, or possibly an acceptance that it is cheaper to pay outright for the land than to contest local property claims. In addition, some of the land acquisition efforts were conducted using measures like bringing along government officials or military personnel to persuade local farmers to go along with the "proposed" transactions (de Jong 1997). In some areas of East Kalimantan the establishment of oil palm plantations has already led to displacement of large numbers of people and the destruction of rattan gardens (Belcher *et al.* 2000). More recent efforts to continue these actions have led to serious conflicts. Reports mention the highly coercive actions of companies like P.T. London Sumatra, which include malicious destruction of rattan gardens and forest by the company. The locals retaliated to these actions by setting to vehicles and buildings and uprooting oil palm trees (Belcher *et al.* 2000).

The coercive occupation of land previously used for FP production was still largely a relic of the pre-*reformasi* period. However, it cannot be ruled out that similar practices will continue for several years at the provincial and district levels because of the new powers lower level state officials now hold. A more structural problem that may seriously affect primary producers' future earnings from FP is the way oil palm is favored and promoted by these officials.

Growing oil palm and related cash crops is attractive to local FP producers because these crops can be incorporated into a larger suite of activities that provides more security of income than is possible with narrow economic strategies. Oil palm companies and regional officials, on the other hand, prefer to exclusively promote widespread oil palm production and eliminate any other land use activity. The persistence of rattan trade barriers continues to artificially reduce the domestic price of

Photo 4-2. Forest products are embedded in local culture and represent local values. Declining local property rights over these resources has impacts that go beyond economic losses. *Photo by Wil de Jong*

the raw material. This includes *illegal* fees charged to traders and official export taxes (Belcher *et al.* 2000). This conjunction of factors leads to the ironic situation in which there is a growing international demand for rattan but large numbers of producers in Kalimantan are abandoning their rattan gardens to join government-sponsored or private oil palm projects. Potter and Lee (1998) have pointed out the mixed experiences of former swidden cultivators who joined smallholder oil-palm production schemes, becoming specialists where they used to subsist on a combination of agriculture, FP collection, and other cash generating activities. The risk of worldwide overproduction of oil palm, and possible price declines, put these previously diversified producers at great risk.

The Conservation of Injustice

The previous sections suggested that even under the new conditions of democratization and related trends of returning the forest to the people, coercion to capture a higher share of profits from FP trade, or to get hold of the land where producers would collect these FP, is likely to persist. Unfortunately, it is not easy to avoid mixing up of agendas, even where true "devolved" management of forests and the resources they contain is pursued. In many parts of the world, as in Indonesia, conservation groups have been at the forefront of efforts to strengthen local control of forests and their resources. Some initiatives have tried to return local forest management to conditions that existed before the outside world gained interest in local resources. Some of these devolution efforts have been somewhat controversial.

In Kalimantan, Kenyah and Merap village aristocrats previously maintained control over the *tanah ulen* (Momberg *et al.* 2000), or forests with high densities of important resources. These privileges were largely abandoned after Indonesia regained its independence from the Dutch. In some recent efforts to achieve forest conservation through increasing communal responsibility of protected areas, Kenyah aristocrats request revival of "local institutions" in the form of *tanah ulen* forest reserves. Conservation activists have presented these efforts as local concern for the environment and community-based conservation actions (for example, Momberg *et al.* 2000). Others have questioned the motives behind these initiatives. Sellato (cited in Vayda 1997) argues that these aristocrats are largely trying to regain control of *tanah ulen.*

Villagers in the Kayan Mentarang district, East Kalimantan, use similar conservation arguments to obtain exclusive rights to natural stands of *gaharu.* Initially these steps started with demands for user fees and shares of the harvest when outsiders came to extract within village boundaries. Eventually these efforts developed into exclusion of all outsiders or confiscation of their collections and equipment. The villagers, with the support of conservation activists, argued that they were protecting resources from over-harvesting (Momberg *et al.* 2000). Vayda (1997), however, argues that there is little evidence of over-harvesting to the point of local extinction. A better explanation is that these actions are efforts by local communities to keep the incomes from *gaharu* for themselves. Similar actions have occurred in other Dayak villages (Appell 1985, cited in Vayda 1997).

Categorizing the Culprits

The True Villains Identified

Using the categories of rent-seeking actions as presented under section two, Table 4-1 suggests a shift in the type of actions that FP producers and traders engage in to obtain larger shares of the profits. True coercive measures in this table persist throughout the history of FP trade up till the *orde baru* period, during which time they are less likely to occur. The examples in Table 4-1 show those to engage in coercive actions to be the sultans, traders, primary producers, the Dutch, the Military, and large companies. It is likely that in the old days there were corrupt lower rank officials who abused weak implementation of rules of the sultans or kings who employed them to capture an extra share of the profits of FP trade. But this type of rent seeking was especially common during the *orde baru* period, and persists to some extent until today. However, during the *reformasi* period this may be caused by a lack of legitimacy of state authorities at different administrative levels, rather than a lack of enforcement or a strategy of central government to assure loyalty to the regime. A second cause of the persistence of this kind of rent seeking during the *reformasi* era is likely the widespread confusion caused by the widely implemented decentralization in the country.

There are several possible explanations for the shift that can be observed in Table 4-1. During pre-colonial and colonial periods collusion was easily possible because traders could, for instance, agree on territorial division in Kalimantan and create monopsony conditions. This was most likely a result of lack of competition as a result of the remoteness of the trading areas, and difficulties to obtain knowledge and trading

Table 4-1. Categorisation of rent seeking actions used by fp producers and traders

Category of action	Pre-Colonial	Colonial	Orde Baru	Reformasi
Those with power engage in coercion	Sultans forced inland groups to collect. Raiding by Taosug. Kenyah, Taosug and Bugis used slaves. Kings claimed share of sandalwood profits.	Kenyah drove away weaker groups from birds-nests caves. Dutch took control of FP trade. Dutch monopolised trade in sandalwood.	Taking control of gaharu trade by PT Saguaro (military). Timber and oil palm companies force people off their land.	
Traders engage in cooperation and collusion	Marriage and blood brotherhood between kings, traders and upland collectors. Bugis suppressed prices for FP by withholding salt deliveries.	Aristocrats control resources or resource rich forests.	Military helicopters used in gaharu collection. Collusion between traders.	Villagers keep competing collectors out of gaharu territories
Those close to central government influence policies or legislation.			NTFP concessions abandoned in the law. ASMINDO is given control of lampit trade. Timber extraction favored over FP collection.	
Provincial and district officials benefit from inadequate enforcement of rule of law from central government			Forest officials allowed illegal harvesting of sandalwood and captured some of the incomes. District offices charge illegal taxes on rattan trade.	District offices continue to charge inappropriate taxes on rattan. Local officials favour and promote oil palm development versus FP development
Actors benefit from ill defined property rights			Timber companies encroaching into forest areas claimed as property by local groups.	Village elites try to reinstate traditional institutions (for example tanahulen). Kenyah villagers still pay percentage of collected FP to aristocrats.

contacts. The latter made it difficult for new outsiders to enter in this trade. In modern times collusion between traders has persisted, but has become more difficult as there are more traders waiting to step in when FP producers want to sell products to other buyers (Peluso 1992). In addition, economic exchange has become common in even the remoter parts of Indonesia. This has made it more difficult for traders to control extractors through debt arrangements, as they used to do.

Inadequate law enforcement and resulting provincial and district officials abusing their positions to capture benefits is largely a result of the fact that only since modern times the state has actually been able to formulate clearly defined rules related to forest products (for example, Weinstock and Sunito 1989). Rather than turn to collusion with colleagues, in modern times the well-connected traders used state agencies to capture excessive shares of the profits in FP trade. A new phenomenon in modern times is also the conflicting interests not over the same resources, but over the resource base[6] where FP traded by forest dwellers are located. Many of the actors who could influence policy and legislation, or get privileged are Jakarta based large companies with interest for timber of land to produce estate crops or industrial timber.

Ill defined property rights was never an issue until recent times. Indigenous groups fought over resource territories, but they were well aware what the property right condition of these resources was under each of the contesting party's customary rights. Outside traders did not have the means to contest the property rights over forest resources by inland groups. In places like Timor, ownership of sandalwood, and distribution of harvested sandalwood among several parties appeared to have been clear for all parties involved. This has changed since the *orde baru* period. Although one of the most pressing problems still is the rights of local people for their forests versus the rights for companies for the land underneath, the two property rights problems in Table 4-1 that persist until modern times relate to a lack of clarity at the village level, and not between locals and outsiders.

Actors Changing Their Strategies

Table 4-1 also suggests the kind of institutional sphere within which several actors operate. Many of them eventually did turn to coercive actions to increase profits from FP trade, including the regular traders. The latter also turned to collusion. There are probably lots of things going on between traders and FP producers after these periods that may come close or actually can be qualified as collusion. We did, however, not find evidence of wide-spread collusion between traders or practices of influencing of government officials by this group, while there is indication in some of the references (for example Peluso 1992) that the importance of collusion between traders directly has diminished.

The kings, sultans, the Dutch colonial government also used coercive measures to get their share of profits from FP trade. During modern times, the upper class rulers of the state hardly needed FP to enrich them selves, as they had incomes from timber, or other more lucrative opportunities. They could be persuaded, however, by the tycoons to prepare legal instruments, favoring powerful interest groups. Those who are in a position to break the rules, without serious punishment are state officials in provincial or district offices, and some of the political well connected.

Lastly, the most diversified among all the actors are the FP producers themselves. They turned to true coercive action, rent seeking activities, and making use of lack of clarity in property rights to obtain larger shares of the profits of FP production and trade.

Conclusions

The previous discussion leads to some useful conclusions that have applied value. Industries based on FP procurement and trade have been around for a long time and wherever certain products gained widespread international attention, many players tried to capture some part of the spills of this trade. All the parties identified in the Indonesia FP industries turned at some point in history to actions that fall within the definition of rent-seeking as identified at the beginning of this paper. Their actions however, differed and could adequately be classified in different categories.

Different actors had a different set of opportunities to use some actions within this set of rent-seeking actions, depending on the special circumstance of time and place in which each of the actors operates. Single actors adjusted their strategies over time. The nature of rent seeking changed over time. The use of economic regulations has increased in importance over history. New actors operating from the central levels of state power became involved and concentrated an a few more lucrative forest products. Their actions included both the creation of rent, and capturing this rent.

There are some general trends in rent seeking related to FP trade. Those who had the power to do so used coercion to benefit from FP trade throughout the history of this trade in Indonesia. The evidence from the material reviewed in this paper suggests that this practice now is much less possible. Collusion between traders has also been a constant throughout history, and is still being used today. However, there is some indication that increased intensity of trade and competition and increased contact between remote regions and major trade centers has slightly reduced the occurrence of trade. Government officials made use of inappropriate implementation of the rule of law from central government especially during the *orde baru* period. This practice appears to continue into the *reformasi* period, maybe less so because of lack of control from central government, but because of the difficulties that are the result of a transition under the decentralization program currently taking place in Indonesia. Outside actors like timber and oil palm companies claim forest territories because there is no consent on property rights, something which happened since the *orde baru*, but it may be more difficult in the *reformasi* era during which there are serious attempts to restore traditional property rights as the officially recognized rights.

When trying to address the issue of rent-seeking activities in conditions where FP development is pursued, this information has some value. On the one hand, trends that this paper observed are long-term and are much longer than the usual time-span that any development effort would consider for effectively addressing rent seeking activities. However, it is useful to recognize that different actors may turn to different kinds of rent-seeking activities. This should make it possible to address such problems in circumstances when FP development is pursued. Alternatively, this knowledge may indicate when FP development efforts will run against too powerful players in the game and are likely to meet limited success. This knowledge may also provide some guidance where FP based development efforts are more likely to run into those problems and when not.

Acknowledgments

We would like to thank Lye Tuck-Po and Arild Angelsen for providing valuable comments on earlier drafts of this paper. We also like to thank the Japan Center for Area Studies of the National Museum of Ethnology for the opportunity to present this paper at the Symposium: The Political Ecology of Tropical Forests in Asia: Historical Perspectives.

Notes

1 We contrast forest products with timber. The former are any products produced by small holders. Timber, on the other hand, is commercial extraction of wood by companies with substantial capital investment.
2 With production of FP, we mean collection from the wild but also collection from managed forests and forest gardens where in addition to the collection some tending and planting of FP may occur.
3 We define political ecology as a collective name for all intellectual efforts to critically analyze the politically-originated problems of natural resource appropriation and political-economic origins of resource degradation.
4 Martua Sirait. International Center for Reseach on Agroforestry, SE Asia Office, Bogor.
5 Indonesian words for "corruption, collusion and nepotism."
6 With resource base we mean the natural population of the FP, but also its surrounding forest and the land.

Bibliography

Andaya, B., and L.Y. Andaya. 1981. *A history of Malaysia*. MacMillan.
Arnold, M. and M. Ruiz Perez. 1998. The Role of Non Timber Forest Products in Conservation and Development. In E. Wollenberg and A. Ingles, eds., *Incomes from the Forest: Methods for the Development and Conservation of Forest Products for Local Communities*, 17–42. Bogor: CIFOR.
Belcher, B. 1998. A Production-to-Consumption Systems Approach: Lessons from the Bamboo and Rattan Sectors in Asia. In Wollenberg, E., and Ingles, A., eds., *Incomes from the Forest: Methods for the Development and Conservation of Forest Products for Local Communities*. Bogor: Center for International Forestry Research.
Belcher, B. *et al.* (11 authors). 2000. Resilience and Evolution in a Managed NTFP System: Evidence from the Rattan Gardens of Kalimantan. Paper presented at the workshop: Intermediate Management Systems Forest Products, Lofoten, Norway, June.
Bryant, R.L. 1992. Political Ecology: An Emerging Research Agenda in Third World Studies. *Political Geography* 11(1): 12–36.
Casson, A. 2000. The Hesitant Boom: Indonesia's Oil Palm Sub-Sector in an Era of Economic Crisis and Political Change (1997–1999).

CIFOR Occasional Paper. Bogor: Center for International Forestry Research.
de Jong, W. 1997. Developing Swidden Agriculture and the Threat of Biodiversity Loss. *Agriculture, Ecosystems and the Environment* 62: 187–197.
——— 1999. Taking NTFP out of the Forest: Management, Production and Biodiversity Conservation. In *NTFP research in the Tropenbos programme: Results and Perspectives*. Wageningen, the Netherlands: The Tropenbos Foundation.
——— 2000. Micro-differences in Local Resource Management: The Case of Honey in West Kalimantan, Indonesia. *Human Ecology* 28 (4): 631–639.
Dove, M.R. 1993. A Revisionist View of Tropical Deforestation and Development. *Environmental Conservation* 20: 17–24.
——— 1996. So Far from Power, So Near the Forest: A Structural Analysis of Gain and Blame in Tropical Forest Development. In C. Padoch and N.L. Peluso, eds., *Borneo in Transition: People Forests, Conservation, and Development*, 41–58. Oxford: Oxford University Press.
Geary, M. and P. Eaton. 1992. *Borneo: Change and Development*. Oxford: Oxford University Press.
Hardenburg, W.E. 1912. *The Putomayo: The Devil's Paradise*. T. Fisher Unwin, London.
Husain, A.M.M. 1983. The Rehabilitation of Sandalwood and the Trade in Nusa Tenggra Timur, Indonesia. Regional Planning and Preparation of Investment Oriented Projects in NTB-NTT. Direktorat Tata Kota dan Tata Daerah, Direktorat Jenderal Cipta Karya, Departemen Pekerjaan Umum, Jakarta.
Mayer, J.H. 1998. Trees Versus Trees: Institutional Dynamics of Indigenous Agroforestry and Industrial Timber in West Kalimantan. Ph.D. Dissertation, University of California, Berkeley.
Momberg, F., R.K. Puri and T. Jessup. 2000. Exploitation of Gaharu and Forest Conservation: Efforts in the Kayan Mentarang National Park, East Kalimantan, Indonesia. In C. Zerner, ed., *People, Plants and Justice. The politics of Nature Conservation*, 259–284. New York: Columbia University Press.
Neumann, R. 1996. Forest Products Research in Relation to Conservation Policies in Africa. In M. Ruiz Perez and J.E.M Arnold, eds., *Current issues in NTFP Research,* 161–176. Bogor: Center for International Forestry Research.

Ormeling, F.J. 1955. *The Timor Problem*. Djakarta and Groningen: J.B. Wolters.

Peluso, N.L. 1983. Markets and Merchants: The East Kalimantan Forest Product Trade in Historical Perspective. M.A. Thesis, Cornell University.

―――― 1992. The Rattan Trade in East Kalimantan. *Advances in Economic Botany* 9: 115–128. New York Botanical Garden.

Peluso, N.L. and C. Padoch. 1996. Changing Resource Rights in Managed Forests of West Kalimantan. In C. Padoch and N.L. Peluso, eds., *Borneo in Transition: People Forests, Conservation, and Develop-ment*, 121–136. Oxford: Oxford University Press.

Potter, L. and J. Lee. 1998. Tree Planting in Indonesia: Trends, Impacts and Directions. Bogor: Center for International Forestry Research *Occasional Paper* 18.

Rohadi, D., R. Maryani, B. Belcher, M. Ruiz Perez and M. Widyana. 2000. Can Sandalwood in East Nusa Tenggara Survive? Lessons from the Policy Impact on Resource Sustainability. *Sandalwood Research Newsletter* 10.

Sellato, B. 2000. Forest, Resources, and People in Bulungan. Elements for a history of settlement, Trade, and Social Dynamics in Borneo, 1880–2000. A report to CIFOR. Bogor: Center for International Forestry Research.

Tapatab, C. 2000. Pokok-pokok pikiran tentang pengelolaan dan pembudidayaan cendana. Paper presented at: Seminar Nasional Kajian Terhadap Tanaman Cendana Sebagai Komoditi Utama Perekonomian Propinsi Nusa Tenggara Timur Menuju Otonomisasi, Gedung LIPI, Jakarta, 26 June.

Vargas, D.M. 1985. The Interface of Customary and National Land Law in East Kalimantan, Indonesia. Dissertation, Yale University, New Haven.

Vayda, A.P. 1997. Managing Forests and Improving the Livelihoods of Forest-Dependent People. *CIFOR Working Paper* 16. Bogor: Center for International Forestry Research.

Weinstein, B. 1983. *The Amazon Rubberboom: 1850–1920*. Stanford: Stanford University Press.

Weinstock, J.A. and S. Sunito. 1989. Review of Shifting Cultivation in Indonesia. Directorate General of Forest Utilization, Forestry Studies, Field Document II-1, Jakarta.

5
Peat Swamp Forest Development in Indonesia and the Political Ecology of Tropical Forests in Southeast Asia

ABE Ken-ichi

Political Ecology and Tropical Forests: Agendas and Issues

The fate of the tropical forests is at the heart of political ecology (Bryant 1992). Before dealing with forestry issues on a broad scale, or in their entirety, we should recognize that there is no one common forest in Southeast Asia. In Whitmore and Burnham's (1984) classification, for example, fifteen different formations can be identified in the region (using such ecological criteria as climate, hydrological patterns, soil types, and elevation). What distinguishes a forest type is reflected in its potential functions, both social and ecological. Forest ecology is affected by the political-economic regimes under which they are governed (Potter, this volume). Shifts of national boundaries also have major consequences for forest history. Political-economic dynamics, how they are enacted differently in different parts of the region, how they change, and how they affect social uses, all have implications for understanding forest histories and forest futures. With changing human use, so do the biological characteristics change. Emphasizing variations among forests in ecological, economical, political, and historical terms is needed to counteract misleading generalizations.

Numerous studies have demonstrated how economic and business-oriented interests have been a principal cause of forest conversion in Asia (Barr 1998; Broad & Cavanagh 1993; Kummer 1992; Leigh 2000; Ross 1996). In the countries concerned, however, a common position is that natural resource exploitation is justified by the needs of national development: the liquidation of forest resources generates necessary foreign currency for the development of the nation and its people. This position remains contentious but is widely held in circles as important as the World Bank, where senior economists argue that logging and forest conversion are essential steps in the development of national economies or that economic processes, through the actions of the free market, will ensure sustainable forest use and therefore forest survival.

Reduced to its bare essentials, this argument falls down on two, quite contrary positions. On the one hand, many conservationists and to a lesser degree indigenous rights advocates argue that tropical forest destruction largely benefits an elite minority or, at any rate, business enterprises with the right political clout. They point to, among other issues, the "hidden costs" of tropical forest exploitation, such as land and water degradation whose effects are borne mainly by forest-dependent communities. On the other hand, highly vocal national politicians, supported by influential development economists, hold that a nation's natural wealth may be used at the nation's discretion for the sake of its people's well being. For justification, the latter points to the unequal distribution of wealth between most Northern and Southern nations. As such, they argue, conservation concerns are something only rich people can afford. Each of these positions implies a different set of proposals to overcome the problems of forest conversion or degradation: reduced impact logging, cede conservation concessions, or increase emphasis on protected areas on the one hand, versus conditionality of World Bank loans, devolved or co-management of forests on the other hand.

The adoption of the development discourse by some of the major actors behind forest conversion represents a peculiar twist to the foregoing debate. This paper will examine a case study of how this was done at a specific time and place. Debates on tropical deforestation as maintained principally by the international conservation caucus are over-general and of little use in understanding local forest histories (Stott 2001). As this discussion will show, the "national development" rhetoric has serious practical consequences, both for the forests and forest-dependent communities. Why a rhetoric of this nature exists, and how it is used to shore up the power base of political elites, is a good demonstration of the relationships between politics, economics, and forest ecology.

The case study here is peat swamp development in Indonesia. The dangers of peat swamp development were starkly revealed in the 1997–

1998 El-Nino droughts and forest fires (discussed in Glover and Jessup 1999). The political literature on forest exploitation is almost silent on the topic of peat swamp development. As will be discussed below, the special biological characteristics of this type of forest constrain the kinds of human activity possible in it. Therefore, it is not enough just to focus on politics or economics; accurate understanding of how ecology limits social uses is important to the analysis. This paper's focus is how peat swamp development was adopted by the then President of Indonesia, Suharto, for his own political uses and ultimately as part of the effort to sustain his legitimacy, leadership, and power.

Indonesia has substantial areas of peat swamp (see below) and has for many years dedicated substantial efforts towards developing these peat lands for the sake of economic development and meeting its target of maintaining self-sufficiency in rice. Many have questioned the wisdom of peat swamp development. They point to the severe difficulties that arise when converting these lands for agricultural production. Among the reasons why planners ignored such well-founded critiques were the underlying desires by those in power to maintain the images of *"pembangunan"* (development). Arguably, these image legitimised the flow of excessive benefits for self, family, and cronies—benefits that came largely from extraneous sources like lucrative logging and oil concessions, access to bank loans without adequate collateral, toll-road concessions, monopolies over clove trade, tax exemptions to develop a national car industry, to name a few.

Apart from the empirical demonstration of Suharto's use of development discourse—to sustain his power-base at the expense of 2 million ha of peat land forest—this paper also offers a concrete study of a local forest history. To elaborate, section two describes the history of peat lands development in Indonesia, a history that started several decades back and is interwoven with the state program of *transmigrasi* (transmigration). The geographic focus is the province of Riau. Section three then moves to Central Kalimantan; it describes the 2 million ha *Projek Pengembangan Lahan Gambut* ("Project to Develop Peat Lands"), also known as the "Mega Rice Project" (Muhamad 2001) or the peat land project, and its role in Suharto's downfall after 32 years in power. In the final section, this particular case is juxtaposed with the global conservation discourse; the intent is to examine what is missing from the global image of forest destruction.

The History of Peat Swamp Forests in Indonesia

Peat Swamp Forest: No Man's Land

Until the late nineteenth century, scientists thought that peat was found only in the cold latitudes. The discovery of peat in the tropics was, therefore, a major surprise. Unlike temperate peat, which has grass vegetation, tropical peat supports forests with trees that may exceed 50 m in height. This type of peat swamp forest, with its distinct physiognomy, occurs almost exclusively in the insular parts of Southeast Asia: in the lowlands of eastern Sumatra, Sarawak, Brunei, Peninsular Malaysia, southwestern New Guinea, and the southern Philippines.[1] The total extent of peat swamp forest in Southeast Asia is estimated to be 19 million ha, of which most is found on two islands, Borneo with 7.8 million ha and Sumatra with 9.7 million ha. In these islands, peat swamp forests extend nearly one fifth of the total area (for more discussion of peat swamp forest ecology, see Yamada 1997).

Traditionally, peat swamp forests were unfavorable environments for human habitation. Damp heat in the daytime and chills at night commonly cause a mist to occur in the forest. Economically, the forests offered little attraction to pioneers and settlers. A deterrent was the high cost of labor involved in clearing or developing the forests. Only the most profitable commercial yields could ever justify these costs. The only exceptions were the more accessible forests at the river mouths. Historically, utilization of peat swamp forest is believed to have started at the Barito estuary by Banjar settlers. The Indonesian word for "peat," *gambut*, is derived from the place-name of this area. The method of exploiting the forest was through tidal irrigation (called *pasang surut*) by which nutrient-rich river water was drained into newly opened paddy and coconut fields by canals built for this purpose. The *pasang surut* technique was occasionally transferred to other areas of the forest when people emigrated in times of social disturbances, but this was not often. Most of the forests, especially in the deep and inaccessible peat areas, remained uninhabited and unutilized until very recently. Indeed, these forests were little studied by scientists precisely because of the difficulty of moving about in them (Yamada 1997: 77).

Today, however, peat swamp forests are rapidly and widely being cleared and developed without restriction. Large-scale establishment of coconut plantations dominates these activities. Like palm oil, coconut oil is an important export earner and much in demand internationally. In Sumatra, one center of peat swamp development is Riau province (facing Singapore

Photo 5-1. A spontaneous migrant opening a canal in peat swamps using simple tools. Riau, early 1990s.

across the Malacca Straits)[2] This new economic opportunity has been a "pull" factor encouraging a rush of immigrants into the province. Riau occupies a strategic position and has recently been developed through regional programs like those of the SIJORI (Singapore-Johor-Riau Growth

Triangle) project, through which physical infrastructure has been expanded in the province.[3] The recent development of peat swamp has had two phases: a phase of arbitrary development by spontaneous immigrants and planned development by "organized" immigrants.

Arbitrary Development by Organized Immigrants

Since the late 1950s, the peat swamp forest areas have been settled by nearby Malay villagers and migrants from Kalimantan and Sulawesi. These were spontaneous migrations driven primarily by small-scale commercial farmers' need for land. They settled at adequate sites at the shore, on the edge between land and the mangrove forests, and gradually started making their way inland. Ethnically, the largest group comprised the Bugis, with their commercial networks; less numerous were the Banjarese and Javanese. At that time, manpower was the main factor needed for reclamation of uninhabited peat lands. Technologically, conversion of peat forest to coconut plantations is simple but laborious: largely it consists of tree-cutting and -burning, canal-building and coconut-planting. Canal-making is a key necessity for agricultural utilization, in order to drain excess water from the peat. No special or complex techniques were used; only manual labor was required. Thus access to labor was the only criteria of value. Old immigrants recall this early phase as a life characterized by "*kapak satu, pisau satu, parang satu, makan sago*" (one axe, one knife, one machete, and eating sago). Land was valueless until forest was cleared and coconuts planted. Only when the coconut trees began to bear fruit was valueless forest transformed into valuable plantations.

Once a coconut plantation was established, little work remained to be done. Harvesting was done only once or twice every three months, followed by copra processing through smoking. Weeding took up just one or two days every three months. Little time and labor was needed to maintain the plantation. As initial sites became non-productive, immigrants would move to new sites and develop plantations there. In addition, they called relatives or acquaintances from their home or former villages and settlements to join them. In the early days, untouched forests seemed limitless and always sufficient for new land-clearance activities, migrants, and settlers. Figure 5.1 shows the results of the encroachment, by analyzing Landsat TM images from 1989.

Planned Development by Organized Immigrants

While spontaneous migration was occurring, peat swamp forests came to the attention of national planners and developers. In fact, peat swamp

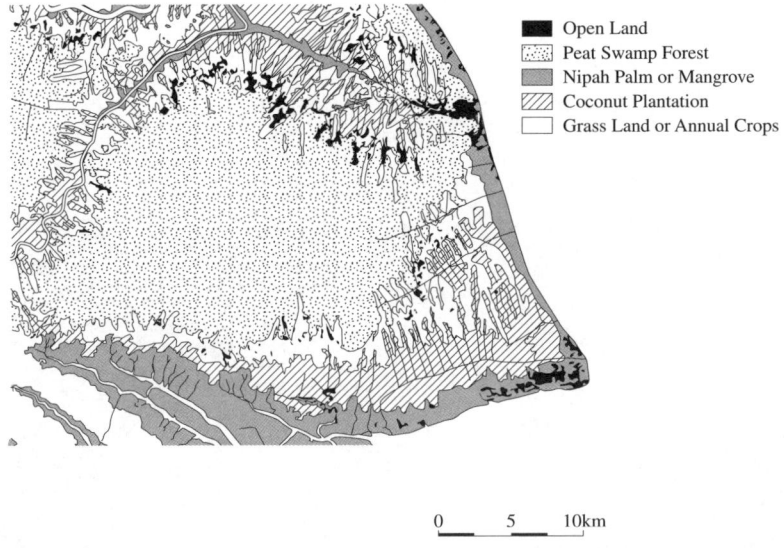

Figure 5-1. Landsat TM images (1989) showing land cleared by spontaneous migrants into peatlands, Riau.

forest clearance was, though strictly illegal, tacitly permitted by local government officials. The migrants' success encouraged government authorities, which then imposed state management of these forests. This resulted in the demarcation of areas for one of the leading state-organized trans-immigration projects, the *Perkebunan Inti Rakyat* (PIR; "Nuclear Estate and Small-holders Project"). PIR projects' stated aims are to increase production of plantation crops and raise the incomes of participating "farmers" as well as to contribute to regional development. The projects recruit personnel from long running transmigration programs. State-owned or private plantation companies support the development of small plantations. The company develops a nuclear estate which it controls entirely, constructs roads, provides houses and other supporting facilities. Participating farmers receive a small area where they are helped to develop a plantation. They are obligated to sell products to the nuclear company at agreed prices.

In the Riau peat swamp area, a PIR project was started in 1991. The project was undertaken by the Sambu Group. This group introduced modern tidal water management techniques, to control for flooding caused by subsidence of the peat land during high tide. The cleared forest area

Photo 5-2. Large canals constructed as a government project, using heavy machinery. Central Kalimantan, late 1990s.

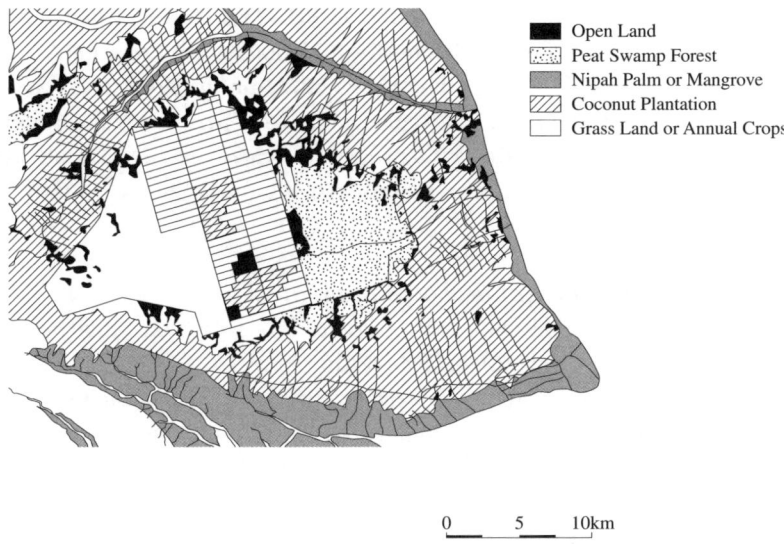

Figure 5-2. Landsat TM images (1997) showing peatland areas in Riau, five years after the operation of the PIR project. The project area is clearly demarcated. Large canals spaced out at regular distances are also visible.

was expanded rapidly as the company employed large construction machinery for forest clearance, canal building and land-preparation. The primary canal is 22 m wide, thus showing a marked difference with the canals dug manually in other locations. The project is entirely planned and arranged, leading to quite a different spatial organization of the developed area. Figure 5.2 shows more or less the same area indicated in Figure 5.1 but five years after the operation of the PIR project. The regular layout of primary and secondary canals is a marked contrast to the *ad hoc* canal arrangements of the spontaneous migrants.

The nuclear estate company strictly controls most of the plantation work and cropping patterns. The spacing of coconut seedlings was initially planned to be 7 m x 8 m but was later widened to 10.4 m x 8 m to allow inter-cropping of pineapple. The coconut seedlings provided by the company are high yield F1 dwarf varieties standardized in the company's laboratory. This variety starts bearing fruit in four years and reaches maximum production after just eight years. The company introduced chemical fertilizers, the use of which it regulates (in terms of frequency of application, quantity and type of fertilizers applied).

By 1997, the area cleared for the Riau PIR-Trans project had reached 110,000 ha, with a total canal length of 25,000 km. Altogether, 12,503 households and 49,902 persons had been settled on the project. Their total populations now account for 11.7 per cent and 9.3 per cent respectively of the prefecture where the PIR-Trans project is located. The impact of such a project is not only ecologically but also socially and economically profound.

Difficulties of Peat Lands Development

Both the structural and chemical characteristics of peat must be taken into account in any agricultural development of these lands. Tropical peat consists mainly of woody materials: the trunks and branches of dead trees. These materials form an underground structure the gaps of which are filled with water and liquefied leaves. The loosely structured peat decomposes when water is drained for cultivation, which then leads to peat compaction and subsidence of the surface. Often, the peat may then become vulnerable to flooding and salt-water intrusion. During the process of subsidence, perennial crops like coconut lose soil from around their roots. The lifetime of coconut plantations planted on peat land is fifteen years at most (Polak 1975), which is less than customary in more favorable environments.

Peat is oligotrophic, lacking mineral nutrients, and is particularly lacking in potassium and phosphorous among the major elements and copper among the minor elements. Therefore, following a certain period of cultivation, as the peat's original nutrients are exhausted, crop yields drop suddenly. Nutrient deficiencies are remarkable in deep peat areas. In the Riau PIR-Trans area, copper deficiency in young hybrid coconuts is reported as early as six to 12 months after planting (Ochis *et al.* 1993). To overcome this problem, the Riau PIR Trans project's nuclear company consumes great quantities of potassium and phosphorous fertilizer. Copper and zinc are imported directly from Malaysia. In effect, coconuts in peat swamp forests are cultivated in a way that resembles the hydroponic method. To reduce the effects of natural compaction, the company uses heavy machinery to pre-compact the ground during the land preparation process. The benefits of pre-compaction are expected to last for twenty-five years but this is by no means certain. The present development of peat is thus potentially risky for the plantation industry and is certainly not sustainable. Spontaneous immigrants have learnt this fact through hard experience. They will, sooner or later, be obliged to abandon their coconut plantations leaving a wasteland behind (Abe 1997).

Table 5-1. Indonesian government statistics showing alleged shortage of rice-growing areas

	1989	1990	1991	1992	1993	1994	1995	1996	1997	1998	1999	2000
Sumatra	2,123,970	2,193,244	2,192,725	2,393,726	1,728,177	2,362,087	2,498,720	2,594,286	2,475,538	2,654,729	2,699,445	2,630,737
Jawa	5,098,892	5,063,461	4,848,079	5,158,968	4,485,670	4,830,643	5,128,225	5,125,689	5,018,453	5,380,268	5,403,429	5,385,138
Bali & Nusa Tenggara	506,324	501,776	499,846	504,228	437,674	503,486	525,367	543,366	524,960	547,650	552,089	555,341
Kalimantan	640,247	657,260	645,889	696,901	460,724	723,003	764,914	782,374	764,617	742,129	887,213	825,818
Sulawesi	991,145	953,019	959,925	1,034,969	695,448	1,056,467	1,147,037	1,185,583	1,084,655	1,125,957	1,217,584	1,018,018
Maluku & Papua	14,378	8,754	14,698	10,315	9,741	18,280	16,957	20,095	13,541	24,829	34,451	27,908
Outer Indonesia sub-total	4,276,064	4,314,053	4,313,083	4,640,139	3,331,764	4,663,323	4,952,995	5,125,704	4,863,311	5,095,294	5,390,782	5,057,822
Total	9,374,956	9,377,514	9,161,162	9,799,107	7,817,434	9,493,966	10,081,220	10,251,393	9,881,764	10,475,562	10,794,211	10,442,960

The Peat Project That Eroded Presidential Credibility

In 1995, President Suharto launched a mega-project in Central Kalimantan to convert one million ha of peat swamp forest into paddy fields. To some extent, recent industrial development in Indonesia had turned to using fertile Javanese lands for manufacturing rather than agriculture. National self-sufficiency in food was achieved in 1985 but was said to become uncertain in the future. The total deficiency of rice for one year was expected to reach three million tons within a few years. The Central Kalimantan peat development project, called PLG (*Projek Pengembangan Lahan Gambut* or "project developing peat swamp area") or the "Mega Rice Project," was primarily intended to make up for rice yield declines in Java.

The threat of a national food deficiency was not supported by the facts from 1995, but it had an appealing political ring. Probably of more importance, justification of the project helped to restrain dissenting voices. Table 5-1, which was reproduced often in the national newspapers, shows Java's alleged shortage of paddy-growing areas to be 12.2 per cent in five years. However, analysts, including those within Indonesia, often consider such government statistics unreliable. In fact, in 1999 government authorities claimed that the rice -growing area in Java had increased to 5,766,614 ha, which is nearly 30 per cent more than in 1993. The growing of food crops is indeed difficult in Indonesia, with most of the fertile areas already taken up, with little opportunity to boost yields in established rice-producing areas, and with problems caused by long dry seasons. However, there have been no obvious deficiencies in the amount of rice produced these past years.

The decision to go ahead with the PLG project was also influenced by the success claimed by the Sambu Group in Riau. This company received the contract to build the infrastructure for the PLG project (Muhamad 2001). The estimated US$ 2 to 3 billion that the project would cost was largely to come from the reforestation fund, with the remaining from the national budget (Muhamad 2001). No external funding was sought for this project.

The necessity and urgency of the PLG project was never questioned openly. Following the official announcement of the project, various authorities expressed their opinions through the mass media, but these never led to a realistic debate, let alone an open critique. Many articles would note the disadvantages and difficulties associated with the project and then conclude along the lines of "We must nevertheless make the project a success." Avoiding controversy had a higher priority than the actual success of the PLG project. This is reflected by the words of Suharto himself who told the consultative committee of the PLG project that "I know all the difficulties. But the essential thing is to start the project."

Sponsoring development projects was a political necessity for Suharto, who was officially given the honorific title of *Bapak Pembangunan* (Father of Development). All the seven Cabinets under his 32 years of presidency were called *Kabinet Pembangunan* (Development Cabinet). Since 1966, Suharto remained head of state under the slogan of Development. In the early part of his regime, development was more than a mere political slogan—Indonesia really did realize growth and "development." Roads and bridges were constructed throughout the country, communication media penetrated into rural areas, transport improved and with it the provision of basic services like health and education. Growth and development was most visible in the skylines of Jakarta, but per capita income also rose steadily from US$260 in 1970 to US$1000 in 1995. The loss of natural resources, forests and forested lands particularly, was compensated for ordinary Indonesians in quantitative improvements to everyday lives. They came to trust in and expect more under Suharto. For Suharto, providing *Pembangunan* also helped to quell social discontent over his family business activities, corruption of his business cronies, suppression of basic political freedoms, and so on.

Continuously presenting new development policies to the nation was therefore a matter of political urgency even before the launch of the PLG project. Public criticism of Suharto's actual development record was suppressed, but it is clear that his reputation was declining. Peat swamp forests, still little utilized, were like the "final frontier" and an easy target for the development impetus. Publicizing a fear of rice deficiency had special appeal to Suharto, who in the mid-1980s had taken personal pride in the national achievement of self-sufficiency in rice. Accordingly, the PLG project not only got financial subsidies (entirely from national resources) but also manpower through the long-running national transmigration program. As planned, one million people would have moved into the project area by 2003.

From the very beginning, the threat of environmental destruction and the economic costs of running the PLG project were obvious to critics. As indicated above, peat land development is potentially unstable and even wasteful and polluting. Furthermore, peat in Kalimantan is relatively shallow and will disappear after only a few years of use. Below the peat layer lies mangrove mud, which, once dried and exposed to the air, turns into an acid sulfate soil generating sulfuric acid. To continue cultivation the toxic sulfuric acid must be washed out via large canals and then neutralized, most economically with lime. Acid water released from the canal affects the aquatic fauna downstream killing economically important fish.

Capital and labor investments in the PLG project are huge. Unlike coconut, which can be sold internationally at high prices, rice cultivation in peatland areas cannot pay for itself. Actually this is why Suharto stressed self-sufficiency of the national staple food as the main project purpose. New F1 rice varieties have been introduced to increase yields, but they inevitably require large quantities of fertilizer and other agro-chemicals. The fertility of peat has become an unavoidable issue, leading to serious discussions about the viability of importing nutrient-rich volcanic ash from Java (*Kompas* 1996: 3).

Participants of projects are initially provided with personal living necessities and also agricultural materials such as rice seeds, a hoe, a sickle, chemical fertilizer, herbicides, and insecticide. However, following the first harvest, new settlers are obliged to buy materials from the core company at fixed prices. The company leases tractors for tillage and threshing machines while it also buys the harvested rice. Participants thus become laborers working for a rice plantation company rather than independent farmers.

Major problems dogged the project. The construction of the infrastructure was not coordinated well. Channels were built but no dams were constructed, which made the canals useless. Warnings issued by the Ministry of Environment that the project would cause tremendous environmental damage were ignored. An Environmental Impact Assessment, legally required for such a project, but only started one year after the project had started, concluded that only about a quarter of the project land was suitable for agriculture (Muhamad 2001). The project was increasingly coming to seem like a major environmental disaster.

On top of these problems came the economic crisis of 1997 which, seen together with the dismal results from the vast expenses outlayed for the PLG project, eroded the prestige of Suharto. Suharto was driven from office on May 21, 1998. In July of that year, the new administration under President Habibie announced its intention to suspend the PLG project.[4] The failed PLG development project contributed to the loss of Suharto's *wahyu*, the divine revelation or a title to be bestowed on the person who is to be a ruler. The *wahyu* for Suharto had been *Pembangunan* (Development). After 32 years in power, the failure of the PLG project demonstrated that Suharto's capacity to further develop the nation through workable and socially appropriate projects had long since waned.

A Political Ecology of Tropical Forest in Southeast Asia?

The Value of Local Histories and Local Forests
The accounts given here are forest histories, a political-ecological history of a specific forest type. It describes the symbolic propulsion of peat swamp forests from peripheral status as an inhabitable and economically "useless" environment into land that will restore a nation's deficiency in critical food supplies. The political use of these forests brought them into the national "commons." This trajectory is not, perhaps, unique in the history of tropical forests. Nor is it the only possible forest history; there are many more, each of them unique, in Southeast Asia.

Fine-grained analyses uncover multiple levels of complexity. Suharto, in his last years in power, had to consider three different constituencies. One included the international community that was becoming increasingly worried over forest exploitation in Indonesia and the consequences that uncontroled development would have on the world at large (see Gale, this volume). The second constituency comprised the President's family members, cronies, and business connections whose wealth came largely from monopolies over timber production and associated industries. As Leigh (2000) shows, the reach and influence of this constituency over local politics and economics is not to be dismissed. The third constituency was the Indonesian people: 200 million citizens who needed to be assured that Suharto was in control of national development.

The PLG project was clearly meant to boost the president's flagging reputation as a sponsor and instigator of development, as explained in the previous section. For the elites, peat swamp development was of little consequence in terms of "business as usual." They already had near-exclusive access to Indonesia's vast timber resources and what little timber was found in Central Kalimantan's peat swamplands was mostly gone already. By 1998, when the project was canceled, almost all of the timber was gone (McCarthy 2001).

Although the PLG development program incensed the international tropical deforestation caucus, the nature of the latter's discourse made it quite easy for a long time to hold off criticism. Tropical forest issues have been popularized worldwide during the past two decades. In this kind of arena, problems like those identified with peat swamp development in Indonesia are resolved through international negotiations, consensus, cooperation, and multilateral treaties (Gale, this volume). However, as the example presented in this paper suggests, global solutions may not be effective. Ultimately world leaders have to deal with the demands of constituencies at home and at home the call for national development has primacy over worldwide demands to preserve tropical forests. As such, conservation demands can and often are (in many producer countries) interpreted convincingly as demands that developing countries sacrifice their rights to growth and progress. Projects like PLG were readily adopted and accepted because they seemed to offer serious development initiatives that would bring important benefits in terms of permanent food security and so on. Even for conservationists, such an argument is unassailable. Demands that countries participate in global standards of conservation may have little practical effect if not accompanied by more sensitive awareness of local-national and national-global linkages. Indeed, active participation in organizations like the ITTO seem to be little more than a smokescreen; at home, the forests continue to be pillaged by ruling elites and the timber establishment.

A new dimension of central importance in the Indonesian context is how "culprits" of deforestation have responded to accusations of misdoing. On many occasions, they have responded by "greening politics." Planting trees or "greening of land" inherently has symbolic power, which can be used to legitimate the power of political authorities; very few people would oppose re-forestation or greening of national lands. "Greening" is one way to get unconditional support from the public of a nation. The way Suharto utilized the peat swamp forests provided a notable example of this in Indonesia. In the process of "developing" (indeed, logging off) the forest, he tried to impress his people by performing the role of a *protector* of the forest. Repeatedly, he appeared on television showing himself planting trees. At the Rio summit, he vigorously played a pro-forest role. Elsewhere in Indonesia, key figures in the timber sector, like the ITCI logging company partly owned by Suharto's close associate Bob Hasan, produced slick television advertisements (televised in and outside Indonesia) suggesting a full awareness of the country's responsibility to manage its vast forest reserves sustainably.

Political Ecology and After

The case study of peat land development demonstrates that the problems of tropical forest destruction are firstly vernacular problems to be solved through the initiative of those most directly connected to the forest. In addition, better tools need to become available to reveal genuine trade-offs between development and conservation. As long as this has not happened there is opportunity for those in power to argue that the projects they propose are important for development, even though they are in fact conducted for other reasons.

In the contemporary era, the nationalization of forests is integral to the process by which globalization proceeds. In the case of peat swamp forests, globalization of the forests occurred first with the migrants' coconut cultivation for the world market and then was nationalized with a state project. This is another example of the common pattern whereby any profitable venture discovered at the local level tends to attract the attention of and be appropriated by political-economic elites (Dove 1993). As forests become encapsulated within national management structures, they come under the purview of the global community, are detached from regional realities and reassembled into a portion of a fictitious global commons. This redefinition of forests is part of the process by which control over the so-called "Southern" forests is firmly placed in the hands of the world.

Political ecology might be an effective approach to analyze the present issues on tropical forests. With this analytical tool, we are able to trace much further root causes and clarify the roles of various forms of power relations in tropical forest management and exploitation. Whether a political ecology approach has the potential to take analysis to the level of action—of changing the balance of power and restoring forest ecology—is, however, a question that needs to be asked.

Notes

1 It also occurs, to a lesser extent, in Central and South America, the Caribbean Islands, and Africa.
2 The province comprises the Riau Archipelago (Kepulauan Riau), covering over 3,200 islands in the South China Sea, and the Riau hinterland (Riau Daratan), located in the mid-eastern part of Sumatra.
3 The project dates from 1989, when the Riau Archipelago became a key part of the "Growth Triangle"—a three-country economic sub-region comprising Singapore, Johor in Malaysia, and Riau.
4 The PLG project was replaced by a new scheme that declares a wider area as an Integrated Development Zone of the Kahayan, Kapuas and Barito Embankment Area.

Bibliography

Abe, K. 1997. *Cari Rezeki, Numpang, Siap*: The Reclamation Process of Peat Swamp Forest in Riau. *Southeast Asian Studies* 34(4): 622–632.

Barr, C.M. 1998. Bob Hasan, the rise of Apkindo, and the Shifting Dynamics of Control in Indonesia's Timber Sector. *Indonesia* 65: 1–36.

Broad, R., and J. Cavanagh. 1993. *Plundering Paradise: The Struggle for the Environment in the Philippines*. Berkeley: University of California Press.

Bryant R.L. 1992. Political Ecology: An Emerging Research Agenda in Third World Studies. *Political Geography* 11(1): 12–36.

Dove, M.R. 1993. A Revisionist View of Tropical Deforestation and Development. *Environmental Conservation* 20: 17–24.

Glover, D., and T. Jessup, eds. 1999. *Indonesia's Fires and Haze: The Cost of Catastrophe*. Singapore: Institute of Southeast Asian Studies.

KOMPAS. 1996. Peat Area will become a National Rice Granary: say Suharto. 16th January 1996.

Kummer, D.M. 1992. *Deforestation in the Postwar Philippines*. Ateneo de Manila University Press.

Leigh, M. 2000. The Political Economy of Tropical Forestry in Sarawak: 1950–1990. Paper presented at the Symposium: The Political Ecology of Tropical Forests in Asia: Historical Perspectives. Japan Center for Area Studies, National Museum of Ethnology. Osaka, November 28–30, 2000.

McCarthy, J.F. 2001. Decentralization of Policy Making and Administration of Policies Affecting Forests in Kapuas District. Internal Report Center for International Forestry Research, Bogor, Indonesia.

Muhamad N.Z. 2001. Management of Tropical Peatlands in Indonesia: Mega Reclamation Project in Central Kalimantan. <http://www.geocities.com/kopitubruk/Report2.html>.

Ochis, R., X. Bonneau and L. Qusairi. 1993. Nutrition Minerale en Cuivre des Cocotiers Hybrides sur Tourbe. *Oleagineux* 48: 65–76.

Polak, B. 1975. Character and Occurrence of Peat Deposits in the Malayan Tropics. In G.J. Bartstra and W.A. Gasparie, eds., *Modern Quarternary Research in Southeast Asia*, 71–81. Rotterdam: A.A.Balkema.

Ross, M.L. 1996. *The Political Economy of Boom and Bust Logging in Indonesia, the Philippines, and East Malaysia, 1950–1994*. Princeton University.

Stott, P. 2001. *Tropical Rain Forest: A Political Ecology of Hegemonic Mythmaking*. IEA Studies on the Environment 15.

Whitmore, T.C. and C.P. Burnham. 1984. *Tropical Rain Forests of the Far East*. Oxford: Oxford University Press.

Yamada, I. 1997. *Tropical Rain Forests of Southeast Asia: A Forest Ecologist's View*. Translated by Peter Hawkes. Monographs of the Center for Southeast Asian Studies English-language series 20. Honolulu: Kyoto University and University of Hawai'i Press.

6
De facto *Decentralization and Community Conflicts in East Kalimantan, Indonesia: Explanations from Local History and Implications for Community Forestry*[1]

Steve Rhee

In 1999 the Government of Indonesia passed a provocative set of decentralization laws. The spirit of Law No. 22 regarding Regional Governance and Law No. 25 regarding Intergovernmental Fiscal Balance in many ways opposes the tradition of centralist control legitimated through previous legislation (Sembiring *et al.* 1999; Down to Earth 2000). Decentralization in Indonesia is a political reaction that attempts to guarantee that regions rich in natural resources will profit from them (Natural Resource Management Program [NRM] 2000) and hence will safeguard against possible secession movements.

Although Indonesia's decentralization laws were passed in May 1999, implementation has been slow, and uncertainty is great among all relevant actors as to how regional autonomy will affect authority over and access to the nation's natural resources.[2] Yet, *de facto* decentralization is taking place—*de facto* because although the level of uncertainty in the policy environment is high, local communities, government, companies, and NGOs in certain areas are strategically maneuvering and positioning themselves, taking actions regarding natural resource use based on their respective understandings (or imaginings) of what decentralization means.

This paper describes how *de facto* decentralization is playing out in a region of the district of Malinau in East Kalimantan that is populated by several Dayak groups and the operational area of a timber company and mining company. Specifically, it discusses how Dayak villagers' actions, predicated upon a new sense of political power, exacerbate conflicts between Dayak groups and within their communities. It also examines how these recent conflicts in the Malinau district are informed by the local history of migrations, interethnic relations among Dayak, and linkages between Dayak groups and powerful outsiders and state policies of territorialization (Vandergeest and Peluso 1995).[3] It also shows how recent conflicts and their historical precedents provide a glimpse of the challenges that community forestry[4] advocates in Indonesia must address in the coming years.

The paper draws conceptually on insights from political ecology, which pays attention to the "politicized environment" (Bryant and Bailey 1997). Proponents of political ecology argue that natural resource issues are to a great extent a manifestation of political-economic forces (Blaikie 1985; Blaikie and Brookfield 1987; Bryant 1992). Blaikie, one of the field's founding figures, argues, for example, that the causes of soil erosion are not proximate or place-based (for example, caused by local peoples alone) but rather are due to inter-linked social and political-economic relations (1985). Accordingly, he suggests, problems of soil erosion will not be solved through "neutral", technical interventions but should involve fundamental changes in political-economic relations. An important contribution of political ecology is illuminating the necessity of linking local natural resource issues with regional, national and global political economy. Through a political ecological lens, this paper provides a case study of the "micro-politics" of resource use decisions and conflicts and the history that informs them. By focusing on these local issues, I attempt to capture the "explanatory roles of different interests, conflict, and social 'agency' among people acting at a local level" (Mayer 1996), thereby opening up the "black boxes" of "local people" and "local communities" that are often counter-posed to equally misleading monolithic portrayals of the "state" in macro-structural explanations (Moore 1993). In doing so, this paper attempts to apply insights from political ecology to provide supporters of community forestry or local forest management with an analysis of a few of the issues that need to be addressed in the coming years in Indonesia.

Following a brief description of the region studied, the paper reviews the history of relations among local Dayak groups and between Dayak and outsiders. It then moves to a discussion of recent inter- and intra-community conflicts, examining them in light of regional history. It then discusses the implications of this study for community forestry initiatives in Indonesia, and concludes with a brief discussion of important issues that will need to be addressed in the decentralization process in Indonesia.

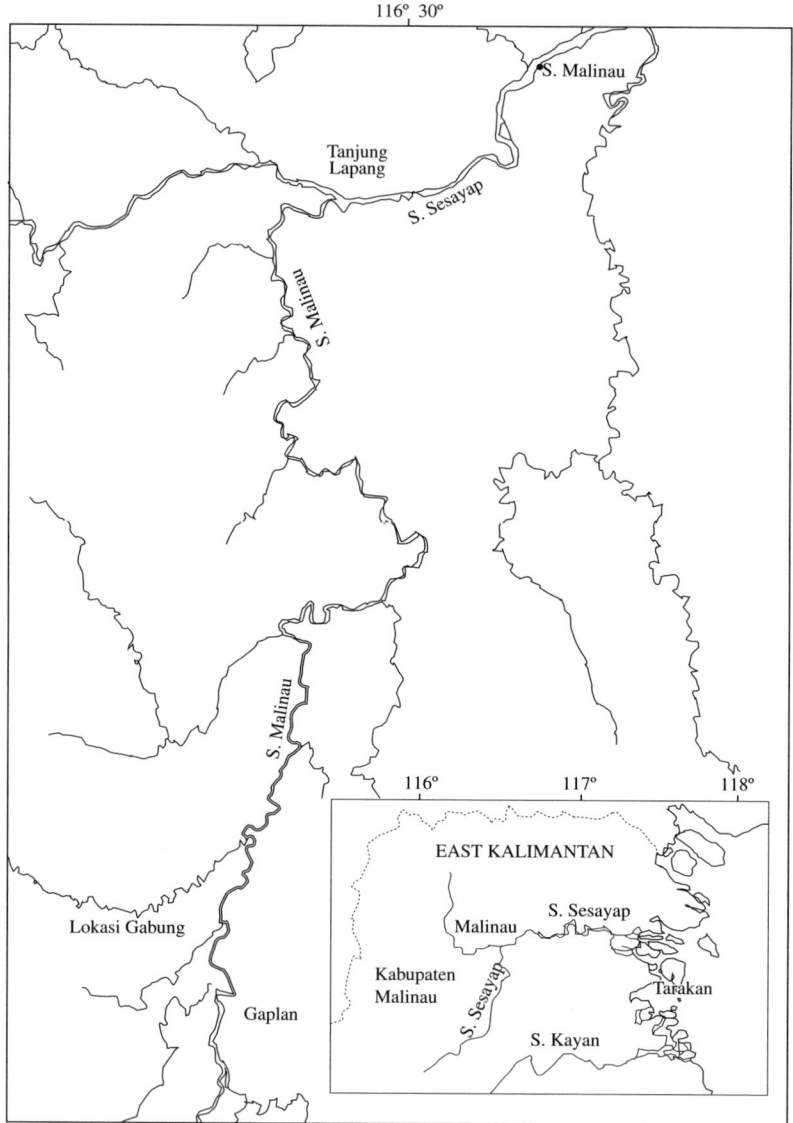

Figure 6-1. District of Malinau, East Kalimantan

Context

The Malinau District of East Kalimantan[5] covers 5,283 km² with a population of approximately 20,000 in 1999 (population density 3.8/km²). Over half of the population lives near the lowland river town of Malinau. Five Dayak ethnic groups—Putuk, Abai, Kenyah, Merap and Punan—live in the interior of this hilly district.

This essay focuses on the Kenyah Lepo' ke, Merap, Punan Malinau and Punan Tubu who live in the four villages collectively referred to as Lokasi Gabung[6] (population 1171). Lokasi Gabung is located upland from the town of Malinau and is situated within a timber concession; it is also adjacent to a mining pit. In 1991 the government awarded a state-owned timber company a 48,300 ha timber concession. Commercially valuable species include meranti (*Shorea* spp.), camphor (*Dryobalanops beccarii*), keruing (*Diptercarpus cornutus*), and agathis (*Agathis borneesis*). A private Indonesian mining company has been extracting coal in the area since 1996. Villagers of Lokasi Gabung travel to Malinau and their swidden fields primarily by hitching rides in mining company trucks, which use the logging road.

Lokasi Gabung consists of four contiguous villages, each village populated by one ethnic group: the Kenyah village of Rehol (population 543), the Merap village of Gayansen (population 250), the Punan Malinau village of Caulan (population 250) and the Punan Tubu village of Labi (population 128). Dayak in these villages have all, at least nominally, converted to Christianity, with the few exceptions of Dayak women who convert to Islam because of marriages with Muslim men. The Kenyah, Merap and Punan groups practice swidden agriculture and hunt wild boar, deer and other animals; maintain gardens for vegetables, fruits and cash crops (coffee and cacao); and harvest forest products such as *gaharu*[7] (*Aquilaria* spp.), rattans, forest fruits and timber. In addition to growing the majority of their subsistence crops, for example, rice, sago, and vegetables, villagers sell forest products and purchase goods such as salt, sugar and clothing in the market town. Villagers seek out wage labor opportunities when they seem advantageous and do not disrupt agricultural activities. Few Dayak in this region work for either the mining or timber company because of the low wages and the infrequent hiring of Dayak.[8] Although villagers resent the timber and mining companies because of the scant benefit they receive, they also view companies as opportunities to gain cash compensation for land.

Overview of History of Inter-ethnic Relations and Relations between Dayak and Outsiders

This section traces the history of the Dayak groups now settled in Lokasi Gabung.[9] The scope of the following narrative is limited to the aforementioned ethnic groups and focuses on the political-economic dimensions of interactions among them and between them and outsiders.

Pre-colonial Period and Social Relations between Dayak Groups

Prior to Dutch arrival to the Malinau in the early 20th century this region was characterized by trade relationships between upland Dayak groups and coastal sultanates and petty kingdoms, who in turn traded non-timber forest products (NTFPs) with seafaring merchants, viz., the Taosug and Bugis (Peluso 1983; Warren 1981).[10] During this era, outside forces did not greatly affect Dayak socio-political relations. As in other regions of Indonesian Borneo, two periods mark changes in the political-economic status of the Dayak: the arrival of the Dutch in the early 20th century and their interventions to safeguard the region from conflict to facilitate trade and the opening of the frontier by the Indonesian government beginning in the late 1960s.

Before the eastward migration of the Kenyah beginning in the 1960s, the Merap[11] were the dominant political force in the upper and middle courses of the Malinau River. Originating west of the Malinau, the Merap conquered the uplands of the Malinau and adjacent Tubu River in the second half of the 19th century with the assistance of local Punan groups and subsequently expanded their control downstream along the Malinau. An important factor in the Merap expansion downstream was the control of birds nest caves along tributaries of the Malinau and other valuable non-timber forest products such as rattan, *gaharu*, and *damar*.[12] The Merap traded these products with the Tidung coastal sultanate, who were the politically dominant group in the area, controlling the flow of trade between seafaring and upland groups.

Although these products have been traded in this region since c.1000 AD, it was only beginning in the 1850s that interior groups were placed in a situation by coastal traders where they had to focus intensively on collecting NTFPs (Sellato 2000: 2.4[13]). Due to the stratified social structure of the Merap, village chiefs and war leaders held private control of valuable resources such as birds' nest caves. The current Merap leader, a descendant of the aristocracy, continues to own the birds' nest caves to this day. Around the turn of the century and after several migrations, the chief Merap leader settled in the village of Gaplan, which is the adjacent, upriver village from Lokasi Gabung and is still the home of the current Merap leader (Kaskija 2000). The Merap now living in Gayansen were part of the Merap migration in the latter part of the 19th century, but did not settle in Gaplan, but rather slightly downstream close to the present day location of Gayansen to which they moved in the early 1980s.

The Merap, who are traditionally swidden cultivators, controlled and exploited forest resources through their long-term relationship with the nomadic Punan,[14] a patron-client relationship that was and to a great extent still is characterized by subservience shown toward the Merap (Kaskija 2000; Sellato 2000). The Merap used the nomadic Punan as collectors of NTFPs, as a military force, as guards of birds nest caves and rattan gardens, while also controlling the flow of resources between Punan collectors and downstream traders. In return for their collecting, the Punan received goods they could not obtain themselves such as iron and tobacco (Kaskija 2000).

The patron-client relationship between agriculturalists and nomadic Punan is a common one throughout Borneo, but the equity of these relationships varies depending on the social structure of the agricultural group with whom the Punan are associated (Rousseau 1990; Kaskija 2000; Sellato 1994; Sellato 2000). The stratified Merap chiefs' strategy was "to keep the Punan as 'Punan,' that is, away from farming and busy collecting NTFPs for them" (Sellato 2000: 2.3.3.2). Indeed, the Punan most closely associated with the Merap, now known as Punan Malinau, "often referred to the Merap chiefs as their *kepala desa* (village chief), and were much more dependent on them than other Punan sub-groups" (Sellato 2000: 2.3.3.2). Furthermore, because of the patron-client relationship and the migration of the Punan with the Merap in the 19th century, the Punan Malinau never formally claimed territory in the region, nor were they concerned with the issue (Kaskija 2000). Although the dominance of the Merap over the Punan Malinau has lessened slightly in the last few decades, many Punan Malinau, including those of Caulan in Lokasi Gabung, still remain subservient to the Merap (Kaskija 2000).

The Punan Tubu now living in Labi, who were never under the direct tutelage of the Merap, experienced a different history than the Punan Malinau. The Punan Tubu were affiliated with less stratified agriculturalists, which allowed for intermarriage, assimilation, and more equitable social relations between the agriculturalists and the Punan Tubu (Kaskija 2000: Rousseau 1990). The traditional relationship with less stratified agriculturalists is one factor that importantly affects the present day attitudes of the Punan Tubu' in the Malinau area. "The most outspoken, politically aware and well-educated Punan are to be found within this category" (Kaskija 2000: 10). As discussed later in the paper, the political awareness and assertiveness of the Punan Tubu' play an important role in interethnic conflicts during the current period of *de facto* decentralization.[15]

Dutch Intervention

It was not until the late 19th century that the Dutch government began a systematic program to control economic and political activities in inland East Kalimantan, spurred by the threat of British expansion into Dutch controlled Borneo and increased demand for NTFPs in Europe (rattan and *gutta percha*[16]) and China and West Asia (birds nests and *gaharu*) (Peluso 1983; Lindblad 1988; Black 1985[17]). It was not until the 1920s that the Dutch fully extended their administration in Malinau. In 1919 the Dutch established a post in the river town of Malinau as a base for military expeditions to suppress headhunting and rebellion and to facilitate trade (Sellato 2000). The Dutch interventions in this region, however, did not dislodge Merap political dominance, although the abolition of slavery and the "increased competition from the Chinese and Malay (Bugis) from the early 1900s probably reduced some of the economic advantages of the Merap aristocracy" (Kaskija 2000: 27).[18]

In important ways the Dutch legitimated Merap political control of the uplands of Malinau, while also bringing it under formal administration.[19] For instance, in establishing a list of birds nest caves and their owners, the Dutch sanctioned Merap control of the caves in Malinau (Sellato 2000: 2.3.1.2). Further, the Dutch and Sultan of Bulungan later made the Merap leader the *Kepala Besar Sungai Malinau/Tubu* (Chief Leader of the Malinau/Tubu Rivers), and this leader's son the *Raja Besar Sugai Malinau* (King of the Malinau River) upon his father's death (Kaskija 2000: 31). Moreover, in 1935, the Merap leader of Gaplan with several other chieftains was a member of the Dutch established *Majelis Kerapatan Besar Tanah-tanah Tidung* (Council of the Tidung Lands) (Sellato 2000: 2.3.1.2).

Post-independence Period

As in much of Kalimantan, for the Dayak living in the Malinau region the post-independence period has meant adapting their traditional social structure and livelihood practices to a whole suite of challenges. Government interventions in the interior and the opening up of the frontier to extract the valuable timber and mineral resources have shifted the political-economic terrain at the local level. Most important for the Dayak in the Malinau region have been the effects of the government's resettlement initiatives, the nationalization of forests, and the extension of the central government into the village political structure.

The majority of the Kenyah[20] currently settled in the Malinau district migrated east from the Bahau and Pujungan river basins in the late 1960s and arrived en masse in the early 1970s during a general period of resettlement[21] spurred by government policy. Both the government and migrants saw resettlement in the Malinau area as attractive because of the region's sparse population (Sellato 2000). Prior to the arrival of the Kenyah there seems to have been limited interaction between the Merap and the Kenyah from the Bahau, who are also traditionally swidden agriculturalists and whose traditional social structure is highly stratified.

Until recently, the Kenyah from the Bahau remained relatively isolated, and hence have maintained their highly stratified social structure, demonstrated by the presence of traditional aristocrats in powerful political positions in Kenyah villages: "In the Bahau and Malinau alike, aristocrats have retained to this day their hold on political power, and the fundamental aristocrat vs. commoner distinction still features prominently in village society" (Sellato 2000: 2.3.2.2).

The Kenyah in Rehol, who identify themselves as Kenyah Lepo' ke, moved to Malinau to be closer to the market, schools, and other benefits of "development." Their migration to the Malinau area followed traditional legal procedure: in the late 1960s the Kenyah village chief requested land from the Merap chief in Gaplan, who subsequently agreed since there was little competition for land and population densities were low (Kaskija 2000). This led to the migration of the Kenyah Lepo' ke en masse in 1975 to the area that is now Rehol. A written agreement was signed by the Merap and the Kenyah Lepo' ke, setting the boundary between the two villages, but was later negotiated in less than transparent ways that still affect inter-village relations (Sellato 2000: 2.3.2.2). In return for the land transfer, the Kenyah paid a small gong as tangible proof of the land transfer and agreed to abide by Merap *adat* (customary law).[22] In this instance, the government's role was limited to sanctioning the transfer. Under the government's resettlement programs, the Punan Tubu, Punan Malinau and the Merap in Gayansen settled in the Lokasi Gabung area during the late 1970s and the early 1980s.

The Kenyah have prospered since their arrival in the Malinau: "Strong social structure, leadership, and discipline, and...their sheer numbers and economic dynamism" have allowed them to succeed economically and challenge the political authority of the Merap (Sellato 2000: 2.3.2.2). As discussed in the next section, the Kenyah, and especially the Kenyah elite, of Rehol have been relatively successful in the context of government disenfranchisement of Dayak livelihood practices.

The migrations and resettlement of Dayak since the 1960s have taken place during a period in which the government has delegitimized Dayak traditional land tenure and disenfranchised traditional social structure through government laws, while also setting out on large-scale exploitation of the nation's tropical forests (Barber *et al.* 1994; Mayer 1996). For example, the Basic Forestry Law of 1967 stipulates that state sanctioned forest production initiatives take precedence over customary law systems, and the Village Governance Law of 1979 serves to strengthen the government's authoritarian rule in villages (Down to Earth 2000).

Summary

In summary, since their arrival in the Malinau watershed and until the late 1960s, the Merap were unequivocally the dominant political-economic force in the middle and upper courses of the Malinau River. Their control of resources upstream and of trade with downstream polities and outside traders were well established. The Punan groups closely affiliated with the Merap in patron-client relationships, now referred to as Punan Malinau, collected these products, receiving in return downstream goods such as tobacco and iron. Due to their relationship with the socially stratified Merap, the Punan Malinau have remained relatively subservient to the Merap and others such as the Kenyah. The Punan Tubu, however, were traditionally affiliated with less stratified agriculturalists, which engendered an assertiveness that plays an important role in current developments in the region. The colonial period and the increased demand for NTFPs legitimated Merap control of the middle and upper courses of the Malinau. Since the later 1960s, Merap dominance of the region has waned due the nationalization of forests in the late 1960s and the arrival of the Kenyah en masse in the early 1970s. As in other parts of Indonesia, the nationalization of forests and other territorialization initiatives systematically delegitimized land claims held by indigenous communities and has contributed to the deterioration of traditional social institutions. Further, since their settlement in the area in the early 1970s, the Kenyah have challenged Merap dominance in the region, creating in the 1990s acrimonious feelings and border disputes between the two groups.

The arrival of the state-owned timber company in 1991 and the mining company in 1996, as well as rumors of future development plans for the region, have increased land value, and hence have created new opportunities for villagers to claim compensation for land. As I discuss in the following section, this has fueled both interethnic and intra-community conflict in the area.

De Facto Decentralization and Local Maneuvres

This section describes how *de facto* decentralization is playing out in Lokasi Gabung and how Dayak villagers' actions, predicated upon a new sense of political power and the increased land value since the 1980s, exacerbate existing conflicts between Dayak groups and within their communities. Further, it discusses how the particular history of relations among ethnic groups contributes to an understanding of the current conflicts.

Perceived Political Power among Villagers

Fieldwork conducted in July 2000 indicates that Dayak in this region now strongly believe that due to decentralization their land rights have been returned to them.[23] The often heard phrase was "our land rights have been returned to us" although upon further questioning, it was evident that local people did not have a firm understanding of this legislation, nor have villagers received any official recognition from the government of their land rights—the region remains *hutan negara* (state forest) and the concessionaire leases are still valid.[24] "*Otonomi daerah*" (regional autonomy)[25] was also frequently part of villagers' explanations. The government, however, has not provided villagers with any sort of legal literacy; villagers' understanding of their land rights returning to them was derived from communication among villagers, the efforts of national NGOs[26] and the Center for International Forestry Research (CIFOR),[27] which has been working in the region since 1998, and/or mass media sources. This understanding—irrespective of how accurate or inaccurate—has been instrumental in changing the attitudes of local people regarding companies and government. There is a sense of political power that villagers themselves acknowledge was rarely felt a year ago. Com-plementing this newfound attitude are the actions and in-actions of the companies currently operating in the area, as well as prospective investors, of which there are many. Indeed the word "investor" itself has seeped into villagers' everyday vocabulary.

For example, village leaders of the Punan Tubu village Labi recently brokered an agreement with an oil palm company to relinquish an area of primary forest that the village leaders claim as belonging to the village of Labi,[28] yet has been granted to a timber concessionaire since the 1980s. The oil palm company negotiated directly with village leaders, ostensibly has the proper permits from the government to operate there, and was building skid trails at the time of the field visit. In speaking with village leaders about this oil palm company, the timber concessionaire never entered into conversations, and villagers' answers to questions regarding the timber concessionaire indicate that the company had not approached them about this situation.[29] In the Merap village of Gaplan, a few km upriver from Lokasi Gabung, a similar incident occurred, but in that case Merap leaders sought out a company to harvest timber in an area already granted to a timber concessionaire.

That these situations exist indicates the impact of *de facto* decentralization. Prior to this period, it was infrequent that villagers in the region possessed the confidence to act so brazenly; that a potential investor would feel the need to meet with local communities, let alone sign an agreement with village leaders to share revenues; and that the existing timber concessionaire would allow this event to take place. This is not to say, however, that prior to decentralization these features were non-existent, but it does seem that the present situation represents a greater recognition of the significance of communities.

These examples indicate that villagers now perceive themselves as an important enough political factor for the companies to consider them relevant to their operations. It indicates a perceived increased bargaining power on the part of local communities. It must be noted that this is a *perceived* sense of political power and may not actually be the case. Indeed, there have been reports from other villages of threats from investors and sitings of surveyors who have not requested permission from villagers (Eva Wallenberg, pers. com.).

Increased Inter-ethnic Conflicts

The new found sense of political power among villagers fuels and exacerbates existing conflicts among Dayak groups, which are predicated upon a particular history of relations among them and with outsiders. For example, in the case of the oil palm company, the primary forest in question is actually claimed by all four villages of Lokasi Gabung. In the past, the villages of Lokasi Gabung had mutually decided that the area in question would be a forest reserve that would be managed collectively, for example, for future swiddens, and hence, individual village boundaries within the forest were not delineated. Because the Punan Tubu have swiddens closest to the area in question, the oil palm company negotiated with their community leaders, who in return did not feel the need to inform the other three villages nor acknowledge to the company that the forest was shared by all four.[30] Moreover, the Punan Tubu village leaders who negotiated the deal were fully aware that their agreement with the oil palm company would breach the previous agreement with the other villages and would spark inter-Dayak conflict.

Once confronted by the community leaders of the other three villages, the Punan Tubu leaders demanded that the only way to resolve the situation was to divide the forest reserve evenly among the four villages, such that each village could make its own decisions about how to utilize the forest. The Kenyah village leaders of Rehol, the village with the largest population (approximately 60 per cent of the entire population of the four villages), strongly disagreed, insisting that the forest be maintained as a collective or that the forest be divided proportionally according to population. Each of the other two villages took opposing sides, and none of the four villages was willing to compromise their individual positions.

The particular history of relations with other ethnic groups explains in part the Punan Tubu village leaders' decision to enter into a contract with the oil palm company. For example, as discussed earlier, their traditional affiliation with less stratified agriculturalists has contributed to the development of political assertiveness vis-à-vis other Dayak groups and outsiders, and the Punan Tubu "are [today] increasingly involved in a struggle for attention and space" (Kaskija 2000: 46). The Punan Tubu also demonstrated their awareness and assertiveness by forming in the early 1990s *Yayasan Adat Punan* (Foundation for Punan Customary Law), which purports to represent all Punan groups in East Kalimantan to preserve Punan culture, indigenous knowledge of the forest, and address violations of *hukum adat* (customary law).[31]

The actions of Punan Tubu village leaders are also influenced by the history of interethnic relations within the four villages of Lokasi Gabung, within which the Punan Tubu, as well as the Punan Malinau and Merap, have been marginalized by the Kenyah. As mentioned earlier, since their arrival to the Malinau in the early 1970s, the Kenyah generally have prospered both economically and politically. The Kenyah in Rehol are no exception. The Kenyah are politically the most powerful group in the four villages. The former long-standing village head of Rehol maintained a seat in the *Dewan Perwakilan Rakyat Tingkat II* (Regency Level General Assembly) prior to the elections in June 1999. A tour through the four villages highlights the disparity in wealth among the ethnic groups and in the services provided by the government and companies. Facilities such as the health post, public meeting hall, primary school, secondary school and village cooperative are all located in Rehol. Moreover, although the timber and mining companies are aware that the Punan and Merap live in Lokasi Gabung, these company officials deal primarily with Kenyah village leaders. These individuals' access to powerful outsiders has brought benefits to the Kenyah, while marginalizing the other Dayak groups.

In addition to this differential access to powerful outsiders, the Kenyah possess a condescending attitude toward the Merap and Punan. Kenyah villagers frequently refer to the Punan and Merap in disparaging terms, often deploying the same rhetoric used by outsiders when they refer to Dayak generally. Many Kenyah speak of Punan, and to a lesser extent Merap, as *terasing* (isolated) and *terbelakang* (backwards) and make explicit that the government relocated the Punan and Merap to Lokasi Gabung so that they could learn from the Kenyah, who characterize themselves as *maju* (advanced). Further, the Kenyah often characterize Punan as having poor hygiene and being lazy. Kenyah villagers also make the point that the Punan do not grow enough rice for themselves and that the rice they do harvest, they sell, and then later must purchase rice at high prices when rice is in higher demand. For the Kenyah, who pride themselves as rice farmers, this translates into not only ineptitude but also stupidity.[32]

Relations between the Kenyah and Merap are more hospitable than between the Kenyah and Punan, yet it is not uncommon for Kenyah villagers to refer to the Punan and Merap as the same lot or to comment disparagingly that the Merap intermarry with non-Dayak, whereas the Kenyah do not. Merap and Punan villagers, however, do not characterize the Kenyah in these terms, but more often comment that the Kenyah are aggressive land grabbers and selfish. In the past, the Merap and both Punan groups expressed their unhappiness with the political dominance and cultural condescension of the Kenyah by establishing their own churches.

Thus, during this period of *de facto* decentralization, inter-Dayak conflicts have been exacerbated by a growing number of powerful outsiders that provide increased, yet differential access to the largesse of "development." Moreover, the precise forms that interethnic conflicts take is informed by the history of relations among Dayak ethnic groups.

Increased Intra-community Conflict

The impact of *de facto* decentralization also exacerbates tensions between village leaders and their constituents. Many villagers feel that village leaders are not representing their best interests in negotiations with companies, but rather are acting in self-interest. Although this sentiment existed prior to *de facto* decentralization, during the July 2000 field visit, villagers more frequently and strongly remarked on the increasing self-interest of village leaders. For example, since the mining company moved into the area in 1996 and severely damaged the river that was a primary source of drinking water for Lokasi Gabung, villagers have demanded that the company build a piped water system sufficient for all four villages. After numerous promises and less than sincere implementation by the company, in May 2000 the villagers renewed their demands for the water system, as well as daily transportation to their rice fields and a network of public lights and requisite generators, both of which had also been informally negotiated and promised in the past. These renewed and more vehement demands resulted in a written agreement between the mining company and the community leaders from the four villages signed by the *Bupati* (Head of the regency government). The timing, intensity and legitimization of the communities' demands indicate that *de facto* decentralization, and the subsequent sense of political power that it has imparted to local communities, played a significant role in the negotiations.

The written agreement, however, differs in significant ways from what was agreed upon verbally during a public meeting, for example, the timeline for the construction of the water system. During the public forum, the company acquiesced to a one-month time line after being refused a three-month deadline. Upon arriving at the Regent's office, however, the company explained to the Regent and the village leaders that one month was insufficient time; the village leaders agreed to provide a three-month deadline. They, however, did not publicly announce this to villagers, and hence when one month passed, villagers were ready to protest. At this juncture, the village head of Rehol explained that the deadline was actually three months. It was this event and others that triggered increased expressions of villagers' suspicion that village leaders were colluding with companies.

The more vocal *expressions* of mistrust of and accusations toward village leaders is a consequence of *de facto* decentralization, i.e., a new sense of political power among villagers. Villagers' mistrust and suspicion of their leaders, however, is to a great extent a result of historical factors regarding the relationship between village leaders and their constituents. For example, with Law No. 5/1979 on Village Governance, the central government enforced a uniform administration system on all villages throughout Indonesia, transforming the role of the village head from a representative of villagers to a representative of the state at the village level, thereby dismantling local forms of order and regulation (Sellato 2000; Down to Earth 2000; Li 1999).

In addition to the government's formal attempt to control village leadership, there is also the issue of the complicity of local leaders with the state: "The position of 'indigenous' leaders may also be complex. Some become brokers of land to outsiders and may betray the interests of their kin and co-villagers in favor of building alliances with newcomers and with patrons promising access to jobs, resources or state power" (Li 1999: 19). In Lokasi Gabung, since the arrival of timber and mining companies, the Kenyah village leaders, especially the former village head, have acted more as agents of the companies than as village representatives. Many villagers are certain that the former village head of Rehol received a salary from both the timber and mining companies. In 1996 when the mining company wanted to mine under villagers' fallow fields, the company dealt primarily with the former village head of Rehol, who brokered a deal with individual farmers,[33] pressuring them to accept half the price they initially requested. Further, Rehol village leaders frequently comment that the mining company "pities" (*kasihan*) local people and wants to assist them, thereby contributing to the notion of the company as a benign patron and villagers themselves as ungrateful clients.

With respect to traditional social structure among socially stratified Dayak such as the Kenyah, the question must be asked regarding the equity and accountability of leaders to their constituents. Rousseau (1990) comments that with Dayak groups such as the Kenyah, "economic and political inequality is buttressed by an ideology which presents stratification as a system of natural (and supernatural) difference" (202). Although traditional social structure provided limited ways for non-aristocratic villagers to express dissatisfaction and transfer allegiances in cases of extreme abuses of power (Rousseau 1990), it is clear that to a great extent inequality characterized the aristocrat-commoner relationship. With the advent of the central government extending its presence in village politics, even these limited means of downward accountability were eliminated.

Implications of this Study for Community Forestry Initiatives

The following section discusses the implications of this study for community forestry initiatives in Indonesia that promote the devolution of authority of forests to local people. Specifically, the case study of the micro-politics of Lokasi Gabung and the history that informs recent local conflicts indicate that in the post-Suharto era and the ostensible transition toward decentralization, advocates of community forestry face complex challenges at the local level that were previously de-emphasized and/or ignored.

Advocates of community forestry in Indonesia—international and national environmental and human rights organizations—often constitute their argument for devolving authority of the forest to local people based on simplified representations of "traditional" communities opposing the "state" to preserve local institutions and practices of sustainable natural resource use (Li 1999). Moreover, within the "state" versus "community" dichotomy, community forestry advocates to a great extent rely upon essentialized representations of upland indigenous or forest-based communities as primitive environmentalists and/or "endangered cultures or tribal people" victimized by the penetration of capitalism (Li 1999). Dayak have been no exception in being located within this rubric by community forestry advocates (for examples see Lynch and Talbott 1995; Poffenberger and McGean 1993).

These representations contributed to recent changes in Indonesian legislation that move toward the devolving of authority of forests to local people. Although in the post-Suharto era these changes open up real possibilities for community-based forestry, they also create complex challenges at the local level that were previously de-emphasized for politically strategic purposes and/or veiled by romanticized notions of forest dependent communities. The case study of recent conflicts in Lokasi Gabung and the local history that informs them illustrates several of these challenges.

The history of settlement, migrations and territorialization initiatives in this region has created a landscape of blurred, overlapping boundaries explicitly contested by multiple Dayak groups. Aware of the increased economic advantages of claiming land rights, the Merap now reclaim the upper and middle courses of the Malinau as their *wilayah adat* (traditional territory), justifying their assertions based on their settlement of the region and on documentation and maps demarcating their territory (Kaskija 2000). Previously the unequivocal dominant political-economic force in the middle and upper Malinau River, the Merap now assert claims over their traditional territory lost through government's initiatives such as the Basic Forestry Law of 1967. However, the Punan Tubu now also assert that the Punan are the original inhabitants of the Malinau and Tubu Rivers, and hence have rightful claim to the region (Kaskija 2000). The establishment of *Yayasan Adat Punan* (Foundation for Punan Customary Law) and its support by an influential national indigenous rights NGO have assisted the Punan Tubu in legitimizing these claims. Overlapping these broader land claims by the Merap and Punan Tubu are the existing land claims of individuals of all Dayak groups settled in the region. Hence, for community forestry advocates the overlapping claims to territory such as those in the Malinau region present a complex problem that has yet to receive adequate attention by practitioners (Campbell 1999), but will be critical to moving community forestry initiatives forward.

Also, this case study challenges the representation of indigenous or forest-based communities as primitive environmentalists or "ecologically noble savages" (Redford 1990), possessing an indigenous environmentalist ethic. Although it was previously advantageous for community forestry advocates to deploy essentialized representations of forest-based communities, recent actions by Dayak in Lokasi Gabung belie these representations and indicate that they look forward to certain future developments and strategically plan for them. The disjuncture between these representations and the actions of forest-based comm-unities provides an opportunity for government to reject previous justifications for community forestry initiatives based on simplistic representations of forest-based communities as primitive environ-mentalists or as possessing intact *adat* (customary law) systems that ensure sustainable use.

Moreover, the traditionally stratified social structure of Dayak groups such as the Kenyah and Merap that are an integral part of their customary law (*adat*) brings into question the equity of these traditional systems—an issue which has received little discussion among practitioners (Campbell 1999). The case study of Lokasi Gabung also reveals the need for community forestry advocates to examine more closely the representativeness and downward accountability of local leaders, as the example of the Kenyah leaders in Rehol indicates.

Conclusion

Through this case study, I have argued that the "micropolitics" of natural resource use in Lokasi Gabung and the history that informs it indicate a more complex reality than expressed by community forestry advocates. Further, it points to challenges that community forestry advocates will need to address in the coming years.

I would, however, emphasize that the details of this case study are at best a "snapshot" of the consequences of *de facto* decentralization; it illustrates one possible direction of change of the socio-political and natural environments in the context of an ambiguous, transitional policy environment. Moreover, it is not the intention of this case study to represent local communities as politically powerful exploiters. At best, the present situation in this area of East Kalimantan shows that the repertoire of the "weapons of the weak" (Scott 1985) has, at least momentarily, broadened.

The effects of *de facto* decentralization on local communities are not entirely positive: this case study has shown that although *de facto* decentralization has seemingly provided opportunities to communities, companies have also received opportunities during this transition period. Furthermore, because of the historical disparity in political power between villagers and companies, there is a strong possibility that regional autonomy will benefit companies and local elite more than the local communities they ostensibly represent if the legislation on regional autonomy is not fully clarified to local people, for example the rights and obligations of social actors regarding the management of natural resources.

Moreover, even if the regional autonomy legislation is sufficiently clarified, certain basic issues in the decentralization process require attention to ensure increased equity, efficiency, participation and government responsiveness, all of which are often promised in decentralization initiatives but rarely occur (Agrawal and Ribot 2000), as this case study indicates. Agrawal and Ribot's (2000) analysis of decentralization of natural resources to local people in Africa and South Asia points to the centrality of downward accountability of local decision makers to their constituents. Although electoral processes are important to operationalize downward accountability there is a need for additional mechanisms. Without such mechanisms for downward accountability, decentralization in Indonesia is unlikely to achieve its ostensible goal. Further, given the history of centralized state control of forest resources in Indonesia and other countries, the move toward effective decentralization will require attention away from conceptualizing "communities as territorially fixed, small and homogenous," which has been the focus of community-based natural resource management advocates (Agrawal and Gibson 1999: 636). And toward a focus on "divergent interests of multiple actors within communities, the interactions or politics through which these interests emerge and different actors interact with each other, and the institutions[34] that influence the outcomes of political processes" (Agrawal and Gibson 1999: 640).

Acknowledgments

The research that informs this essay was conducted in collaboration with the Center for International Forestry Research (CIFOR) from June-September 1999 and July 2000. Fieldwork for the case study was carried out in a region of East Kalimantan, a long-term CIFOR research site. I thank Lini Wollenberg for comments on portions of an earlier draft and also for her unfailing encouragement. I also thank the field staff, specifically, Njau Anau, Asung Uluk, Made Sudana, and Godwin Limberg, for sharing their insights on and facilitating my understanding of the socio-political dynamics of this region. I am indebted to the villagers of Lokasi Gabung for their hospitality, unending patience with my questions, and willingness to allow me to participate in their lives. CIFOR not only made my research in this region possible but also provided me with full logistical support. I also thank Bernard Sellato and Lars Kaskija for allowing me to use their unpublished reports on the region. I received generous financial support for the research in 2000 from the Council on Southeast Asian Studies at Yale University and in 1999 from the Council on Southeast Asian Studies, the Program in Agrarian Studies and the Tropical Resources Institute at Yale University. I claim full responsibility for any shortcomings and omissions.

Notes

1 This paper is based on fieldwork carried out in Indonesia from June-September 1999 and July 2000.
2 The levels of uncertainty and disparity in perceptions regarding decentralization and its implications were captured in a study carried out by the Institutional Task Force for Forestry Sector Decentralization under the auspices of the Ministry of Forestry and Estate Crops (MFEC 2000).

In this study, the task force conducted interviews and group discussions with central and regional representatives of government agencies, legislative bodies, NGOs, universities, and communities, and held a three-day workshop with regional government representatives from six provinces (MFEC 2000). This study revealed that there is no consensus on what decentralization means, how it should be im-plemented, and how roles and responsibilities are divided. The results of this study also show that central government and regional government officials have significantly different opinions regarding authority over and responsibility for forests (MFEC 2000).

3 I use the term territorialization to mean "the process through which 'all modern states divide their territories into complex overlapping political and economic zones, rearrange people and resources within these units, and create regulations delineating how and by whom these areas can be used'" (Vandergeest and Peluso 1995: 387, cited in Li 1999: 12).

4 "Community forestry" is a diffuse term that has changed in meaning over time (Gilmour and Fisher 1997). Dove (1995) notes that a review of the state of the art in social forestry revealed that there is marked disagreement about its basic tenets. I use the term to refer broadly to those initiatives that promote the devolution of rights over access and control of forest resources to local people.

5 At the end of 1999, the administrative boundaries within the province of East Kalimantan were altered; the government decided that existing boundaries covered too much territory for effective governance. As part of this process, the status of Malinau was "upgraded" from *kecamatan* (district) to *kabupaten* (regency) and the boundaries for the region redrawn. The figures used in this paper reflect those prior to the administrative changes.

6 Village names are pseudonyms.

7 *Gaharu* is the resinous heartwood that results from a fungal infection in some species of *Aquilaria*. The resulting aromatic heartwood is exported and used in perfumes and incense.

8 A sub-contractor began operating in the area in the last year and unlike the mining and timber companies employs a substantial number of Dayak villagers.

9 Recent unpublished reports by Sellato (2000) and Kaskija (2000) provide extensive histories of the region of Malinau based on both local oral histories and archival research. Here I refer to them and draw on ethnographic and secondary historical studies of the broader region and of Dayak groups.

10 Trade in NTFPs with China is recorded in Chinese archives dating to c. 1000 AD (Sellato 2000)
11 Besides reports by Sellato (2000) and Kaskija (1999; 2000), I have not found historical or anthropological material on the Merap.
12 *Damar* is a resin from trees of the *Dipterocarpaceae* family and is used to make matches, varnish, and turpentine (Peluso 1983).
13 The draft of the report I received does not have page numbers, hence I cite specific sections of the report as a proxy for page numbers.
14 For ethnographic studies of the Punan see, for example, Sellato (1994; 2000), Hoffman (1986; 1988), Kaskija (1999; 2000), Puri (1997).
15 Both the Punan Malinau and the Punan Tubu now open swiddens and do not entirely rely upon trade in NTFPs. See Sellato (1994) for an explanation of how Punan became sedenterized.
16 *Gutta percha* is a latex from the trees of the genus *Palaquium*; its most important use was to insulate telegraph wires and cables in submarines (Peluso 1983).
17 Black (1983) notes that the oil fields in Tarakan had little effect on upland political-economic relations.
18 Opinions differ with respect to the extent and character of Dutch involvement in the interior. Whittier (1973: 38) notes that in the Apau Kayan the Dutch were "relatively unobtrusive in terms of demanding changes in the Kenyah lifestyle with the exceptions of headhunting and warfare." Sellato (2000) shares the same sentiment with respect to the Merap in Malinau, de-emphasizing the effects of the elimination of headhunting, warfare and slavery. According to King (1993: 158–159), their elimination resulted in redirecting power and influence to the Dutch, as well as in indigenous leaders looking for "other avenues of advancement and leadership."
19 Rousseau (1990: 198) notes that "colonial rule brought about a stabilization of regional leadership into formally defined political offices under administrative control."
20 For studies of Kenyah social stratification see, for example, Whittier (1973) and Rousseau (1990). The total population of Kenyah in Malinau in 1990 was 1665.
21 The government did not directly force upland people to resettle in this area. Villagers themselves initiated some migrations, while in other cases, the government strongly encouraged resettlements.
22 Kaskija (2000) notes that this was not a transfer of ownership, but rather a granting of use rights. Regardless, all groups moving into

the upper and lower courses of the Malinau acknowledged the Merap's territorial ownership and requested land from them.
23 This may also be due to villagers' nascent awareness of other government regulations such as Regulation from the Minister of Agrarian Affairs/Head of the Bureau of Lands No. 5/1999 Guidelines to Resolve *Adat* (Customary) Communal Rights Conflicts, in which the National Land Agency will accept the registration of *Adat* lands and treat them as a communal and non-transferable right (Fay and Sirait 1999). Villagers in Lokasi Gabung never specifically mentioned this government regulation, and none had their lands registered. Leaders in other villages, however, are aware of the possibility and are attempting to register land as customary.
24 Whether the leases will remain valid after full implementation of decentralization is uncertain.
25 This Indonesian term is used nationally as a synonym for decentralization (*desentralisasi*).
26 A national NGO has worked in the area for the last year, and its work has influenced local people's attitudes.
27 This region is a long-term research site of CIFOR.
28 Prior to the arrival of the oil palm company, the four villages collectively claimed this area of forest. See following discussion.
29 This is not to imply that the timber concessionaire is unaware of the situation or that the concessionaire has not approached the government or oil palm company.
30 Who initially approached whom in this situation is unclear.
31 The actual representativeness of this organization is suspect; informants as well as recent studies by Sellato (2000) and Kaskija (2000) indicate that the organization to a great extent only represents those Punan Tubu actively involved in the organization.
32 Similar to the Merap, the Kenyah have had long-term patron-client relationships with Punan groups in which Kenyah have held a dominant position (Sellato 1994, 2000; Hoffman 1988; Kaskija 1999, 2000).
33 This is not uncommon and has occurred in other villages in the area. Evidence indicates that the mining company bribes the local headman to broker deals to ensure that the community receives marginal compensation.
34 "Institutions" are defined as "sets of formal and informal rules and norms that shape interactions of humans with others and nature" (Agrawal and Gibson 1999: 637).

Bibliography

Agrawal, A. and C. Gibson. 1999. Enchantment and Disenchantment: The Role of Community in Natural Resource Conservation. *World Development* 27(4): 629–649.

Agrawal, A. and J. Ribot. 2000. Accountability in Decentralization: A Framework with South Asian and African Cases. *Developing Areas* 33 (summer): 473–502.

Barber, C., N.C. Johnson and E. Hafild. 1994. *Breaking Through the Logjam: Obstacles to Forest Policy Reform in Indonesia and the United States*. Washington, DC: World Resources Institute.

Black, I. 1983. The 'Lastposten': Eastern Kalimantan and the Dutch in the Nineteenth and Early Twentieth Centuries. *Journal of Southeast Asian Studies* 16: 281–291.

Blaikie, P. 1985. *The Political Economy of Soil Erosion in Developing Countries*. New York: Longman.

Blaikie, P. and H. Brookfield, eds. 1987. *Land Degradation and Society*. London: Methuen.

Bryant, R.L. 1992. Political Ecology: An Emerging Research Agenda in Third World Studies. *Political Geography* 11(1): 12–36.

Bryant, R.L. and S. Bailey. 1997 *Third World Political Ecology*. Routledge: London.

Campbell, J.Y. 1999. Hutan Untuk Rakyat, Masyarakat Adat, atau Kooperasi? Plural Perspectives in the Policy Debate for Community Forestry in Indonesia. Paper presented to the Seminar on Legal Complexity, Natural Resource Management and Social Security in Indonesia, September 6–9, Padang.

Down to Earth. 2000. Special Issue on regional autonomy, communities and natural resources. *Newsletter* 46.

Fay, C. and M. Sirait. 1999. Reforming the Reformists: Challenges to Government Forestry Reform in Post-Suharto Indonesia. Paper presented at the American Association of Rural Sociology, August 6, Chicago, Illinois.

Gilmour, D.A., and R.J. Fisher. 1997. Evolution in Community Forestry: Contesting Forest Resources. Paper presented at the conference on Community Forestry at a Crossroads: Reflections and Future Directions in the Development of Community Forestry, Bangkok, Thailand, July 17–19.

Hoffman, C.L. 1986. *The Punan: Hunters and Gathers of Borneo*. Ann Arbor, Michigan: UMI Research Press.

———— 1988. The 'Wild Punan' of Borneo: A Matter of Economics. In M.R. Dove, ed., *The Real and Imagined Role of Culture in Development*, 89–118. Honolulu: University of Hawaii Press.

Kaskija, L. 1999. Stuck at the Bottom: Opportunity Structures and the Punan Malinau Identity. In B. Sellato and P. Sercombe, eds., *Forest Minorities: Cultural Survival and Social Change among Borneo's Hunter-Gatherers*.

———— 2000. Punan Malinau and the Bulungan Research Forest: A Research Report. Unpublished report for the Center for International Forestry Research, Bogor, Indonesia.

King, V.T. 1993. *The Peoples of Borneo, Peoples of South-East Asia and the Pacific*. Oxford: Blackwell Publishers.

Li, T.M. 1999. Marginality, Power and Production: Analyzing Upland Transformations. In T.M. Li., ed., *Transforming the Indonesian Uplands*, 1–45. The Netherlands: Harwood Academic Publishers.

Lindblad, J.T. 1988. *Between Dayak and Dutch: The Economic History of Southeast Kalimantan 1880–1942*. Dordrecht: Holland Foris Publications.

Lynch, O.J. and K. Talbot. 1995. *Balancing Acts: Community-Based Forest Management and National Law in Asia and the Pacific*. Washington, D.C.: World Resources Institute.

Mayer, J.H. 1996. Trees vs. Trees: Institutional Dynamics of Indigenous Agroforestry and Industrial Timber in West Kalimantan, Indonesia. Ph.D. Dissertation, University of California, Berkeley.

Ministry of Forestry and Estate Crops (MFEC). 2000. *Institutional Task Force for Forestry Sector Decentralization, December 1999–February 2000*. Jakarta, Indonesia: Ministry of Forestry and Estate Crops.

Moore, D.S. 1993. Contesting Terrain in Zimbabwe's Eastern Highlands: Political Ecology, Ethnography, and Peasant Resource Struggles. *Economic Geography* 69 (4): 380–401.

Natural Resources Management Program (NRM). 2000. *NRM News* 1(1). Jakarta, Indonesia: Natural Resources Management Program.

Peluso, N.L. 1983. Markets and Merchants: The Forest Product Trade of East Kalimantan in Historical Perspective. Unpublished MA thesis, Cornell University.

Poffenberger, M. and B. McGean, eds. 1993. *Communities and Forest Management in East Kalimantan: Pathway to Environmental Stability*. Berkeley: University of California Press.

Puri, R.K. 1997. Hunting Knowledge of the Penan Benalui of East Kalimantan, Indonesia. Ph.D. Dissertation, University of Hawaii.

Redford, K. 1990. The Ecologically Noble Savage. *Cultural Survival Quarterly* 15: 46–48.
Rousseau, J. 1990. *Central Borneo: Ethnic Identity and Social Life in a Stratified Society*. Oxford: Clarendon Press.
Scott, J. 1985. *Weapons of the Weak: Everyday Forms of Peasant Resistance*. New Haven: Yale University Press.
Sellato, B. 1994. *Nomads of the Borneo Rainforest: The Economics, Politics, and Ideology of Settling Down*. Honolulu: University of Hawaii Press.
─────── 2000. Forest, Resources, and People in Bulungan: Elements for a History of Settlement, Trade, and Social Dynamics in Borneo, 1880–2000. A report to CIFOR. Bogor: Center for International Forestry Research.
Sembiring, S.N. and F. Husbani. 1999. *Kajian Hukum dan Kebijakan Pengelolaan Kawasan Konservasi di Indonesia: Menuju Pengembangan Desentralisasi dan Peningkatan Peranserta Masyarakat*. Jakarta, Indonesia: Natural Resources Management Program.
Vandergeest, P. and N.L. Peluso. 1995. Territorialization and State Power in Thailand. *Theory and Society* 24: 385–426.
Whittier, H.L. 1973. Social Organization and Symbols of Social Differentiation: An Ethnographic Study of the Kenyah Dayak of East Kalimantan. Ph.D. Dissertation, Michigan State University.
Warren, J.F. 1981. *The Sulu Zone, 1768–1898: The Dynamics of External Trade, Slavery, and Ethnicity in the Transformation of a Southeast Asian Maritime State*. Kent Ridge: Singapore University Press.

7
One Hundred Years of Land-Use Changes: Political, Social, and Economic Influences on an Iban Village in Bakong River Basin, Sarawak, East Malaysia

ICHIKAWA *Masahiro*

This paper describes land-use changes[1] in Nakat, an Iban village located in the Baram River area in northeastern Sarawak, East Malaysia (see Figure 7-1). In this village the forest has been central in the development of the local economy, providing resources for domestic and commercial uses, and land that could be cleared for agricultural production. However, land use has not remained static through time. Extra-village influences, such as local and international market fluctuations, have been critical to the process of change. Even the beginning of the village itself, instigated by settlers' search for primary forest, was ultimately due to government policies promoting cessation of warfare and economic development of the territory. In some periods of time, collection of products from swampland forests was done enthusiastically; at other times, that same forestland was converted into rice fields and rubber gardens. Not only has land use changed, but also the kinds of products that villagers sought to grow and to collect from forests. As extra-village conditions changed, their effects percolated through to the local economy, affecting, for example, methods and techniques of rice cultivation and site selection for farmland. Recently drastic changes in the subsistence base and land use can be linked to road and urban development. It seems that villagers in Nakat have not lived a passive self-sufficient life but have tried to deal positively with, and economically benefit from, changes in external political and social conditions that have affected the village.

The negative impacts of external influences, such as increasing degradation of natural environments and loss of local ability to deal with environmental perturbation, are often emphasized in peasant studies (Watts 1983; Grossman 1981; cf. Donovan, this volume). The resulting resistance of local people to government policies and their implementation have been described in several studies on political ecology (Peluso1992; Watt1983). Such resistance was not observed in Nakat. This may partly be because degradation did not occur in Nakat. An alternative explanation may be that local communities have resilience and flexibility to adapt their local natural resource management to accommodate to forces of change that elsewhere lead to degradation. In Nakat, adaptation to external conditions has been thoroughly integrated into the mechanism of survival in a severe and unstable natural tropical forest environment. This mechanism is observed not only in the study village, but also in the wider areas of Sarawak and insular parts of Southeast Asia.

The villagers of Nakat recognize several phases in the history of their village, each marked by different land-use patterns and economic activities. The shifts from one phase to another were caused by wider political and economic conditions. This paper distinguishes three historical periods: the pioneer period (1900–1950), the period of commercialization of rice and rubber (1950–1985), and the period of diversification of commodities (1985–1998).

The objective of this paper is to show the strong connection between land-use patterns and political, economic and social conditions outside the village. Section one discusses the methods used in the research leading to this paper, followed by an introduction of Nakat in section two. Sections three to five then examine the three periods, related to the major changes in land use. Finally, the paper concludes by interpretating the observed changes in village history in terms of the villagers' flexible responses to changes in political and economic conditions far away from their own homelands.

Methods

I conducted field research for this paper in Nakat during four separate periods: February-June and October-November 1996, July-August 1997, and August-December 1998. During those periods, I stayed with a family in the village and conducted interviews with villagers on land-use histories.

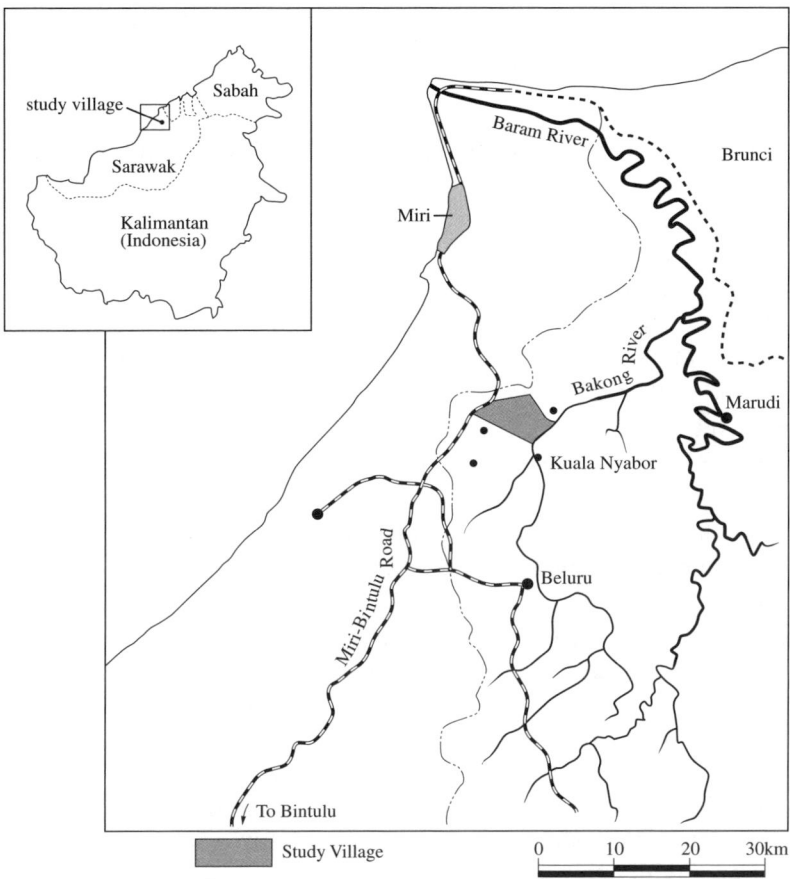

Figure 7-1. Bakong River basin, Sarawak

Observations and measurements of current land-use patterns and activities supplemented these interviews. Chinese merchants of Bakong River basin and extension officers from the Agriculture Department also provided information on changes in resource use. These merchants sometimes have had contact with the villagers coming to their shops since before the Pacific War, and the extension officers visited the village occasionally over the past 20 years. They are therefore familiar with its land use and other economic activities. Textual sources like the *Sarawak Gazette* (SG), which reported on conditions in the Baram area, permitted independent confirmation of events documented through interviews. In addition, I used aerial photographs from 1947, 1963, 1977, and 1997 for a spatial analysis of land-use changes.

Location, History and Natural Social Conditions of Nakat

The village of Nakat lies between the Miri-Bintulu road and Bakong River, which is a tributary of the Baram River. The village's territory of 37 km² is located in the Miri Division, which extends over 27,000 km² in the northeast of Sarawak, and has a population of 233,000 (Department of Statistics, Malaysia 1997). As of May 1996, the village population[2] was 454, almost all of them Iban. Iban are the major ethnic group in the state, comprising one third of its population. They usually live in longhouses containing household compartments called *bilek*. Longhouse communities are egalitarian and have no distinctions in rank, as other groups in Borneo do (see Rhee, this volume). The *bilek* is a socially highly independent unit of the longhouse community (Freeman 1955).

The village of Nakat was founded at the turn of the 19th century. The Brooke[3] government had started the physical and economic development of the Baram area after Brunei ceded this territory in 1882. Initial activities focused on the construction of strongholds, offices, and market towns to trade forest products. The first settlement of the Baram basin dates from 1891, when James Brooke permitted Iban for the first time to move to this region (Pringle 1970: 271–272).[4] Previous to that time no Iban had lived permanently in this region. According to the 1907 census, the number of Iban living in the Baram already had increased to 1,960 or 12 per cent of the total population (SG 4/4/1907: 83).

There are two reasons behind the Brooke settlement program into the Baram region. First, it was expected that new settlers would secure peace and compliance with the rule of law in this frontier area with Brunei. Secondly, the program also intended to expand the collection of forest products. The Iban were considered ideal settlers because Brooke knew their warfare qualities well and felt he could count on them in case of military emergencies (Pringle 1970: 270–271). Forest products such as rattan, forest rubber, and dammar were very important economic resources for Sarawak at the time, regularly accounting for about a third of its total exports between 1876 and 1910. In most of the newly acquired districts trade in forest products was the only potential source of revenue (Pringle 1970: 267).

Iban from the Second and Third Divisions of Sarawak, for their part, were in critical shortage of land (Pringle 1970: 272). They migrated to the Baram area seeking new lands and new forests in which to collect forest products.[5] The village of Nakat was only one of the villages established along the Bakong River. According to oral history, the founders of Nakat split off from a group that settled in the village of Malang, which is located in the vicinity. The split occurred just after Malang was established, which occurred in 1896 (Sandin 1994: 228). Thus the foundation of Nakat can be estimated to be around 1900. During that time six Iban families constructed a longhouse near the mouth of Song River, which flows into the Bakong River. The six families had originally come from the Nyelong River, a tributary of the Rajang River in the Sibu district.

The middle and lower parts of the Bakong River hold abundant swampland. The swampland belonging to Nakat, about half of the village territory, is distributed along the Bakong River. The rest of the village contains hilly lands, land through which the Miri-Bintulu road now passes, but even this hilly part of village territory has valleys where swampland occurs. There is a difference of some 170 m in elevation between hills and swampland along the Bakong River; the latter is less than 15 m a.s.l.

The village has strong economic ties with the city of Miri (Figure 7-1), which developed after the discovery of petroleum in 1910 and has become the third largest city in Sarawak. In 1991, its population was about 130,000 (Department of Statistics, Malaysia 1997). Nakat villagers purchase food and other daily necessities and sell products in Miri. The villagers also have links with other smaller towns and villages, such as Marudi and Beluru. Marudi has been the main town in Baram River and nowadays has about 130 Chinese shops. Beluru serves a similar function for the Bakong River and also has several Chinese shops. Kuala Nyabor had until 1985 four Chinese shops patronized by the nearby Iban villagers but the Chinese have since then left this location.

The Pioneer Period (1900–1950)

In the early years of Nakat's existence the villagers engaged in two main economic activities, the cultivation of rice and secondary crops[6] in Penan primary forestlands, and the collection of commercial forest products. In 1930s, a few villagers made pepper gardens near the longhouse, but the scale of cultivation was small and they were abandoned after the start of the Pacific War (1941–1945). According to oral history, the Bakong area was formerly part of Penan territory. Following the Iban's arrival, the hunter-gatherers reportedly moved to Tinjar River, a tributary of the Baram and more than 30 km away. The *Sarawak Gazette* (1/4/1902: 79) reports, for example, that the Iban at Laong (upstream from Nakat) had settled in Penan territory. Initially, primary forests (*kampong*) occupied most of the land area. The villagers who established a longhouse near Bakong River started to clear land for dry and wet rice fields, first around the longhouse, then gradually expanding outward. Aerial photographs from 1947 confirm oral history that the extent of the primary forest opened up at this time was from the longhouse to a spot near today's Miri-Bintulu road, which is also the village border. On the other hand, unfelled forests still remained in the vicinity of the Bakong River, where floods make agricultural production too risky, and on hilly ridges where the soil is poor and access from the longhouse is difficult (Figure 7-2a).

Clearing primary forests was also a way to claim usufruct rights to the lands. Following Iban customary law, usufruct is given to the one who first fells a primary forest, and the rights are then passed on to his descendants (Freeman 1955). Villagers therefore clear primary forestland partly also to increase landholdings. The villagers preferred to grow hill rice rather than wet rice in swamplands. For one thing, the hillside fields could also sustain the growing of secondary crops, which supplemented the diet at times when these crops could not be obtained through trade. The reason for growing rice in both dry hillside and wet swampland fields was, as also argued by Dove (1985), to minimize risks of poor harvest caused by droughts or floods.

The cultivation of hill rice follows an established sequence. Cutting a patch of hill forest opens a plot of land that is left to dry and then burned. Rice seeds and secondary crops are planted using a dibble stick to make holes in the soil in which a few rice seeds are put. After one to three years of cultivation on the same land, another patch of forest is felled and opened for cultivation. A similar method is used for wet rice cultivation. However, after the swampland forest has been slashed and burned, the rice seeds are broadcast, rather than planted with a dibbling stick. Today, many villagers have adjusted this technique and clear swampland that is under grass vegetation, into which rice is transplanted (Ichikawa 2000a). Broadcasting seeds is more effective, partly because of the many unburned and felled trees scattered around the rice fields, which makes moving around the fields difficult. Wet rice fields usually are left to fallow after a few years of cultivation, thus allowing forest to grow back on the site. Although the pioneer period was characterized by clearing of primary forests, rice fields were not always opened in primary forests, but instead in fallow forests located in easily accessible areas from the longhouse. For example, distribution of wet rice fields opened in fallow forests are shown as *Scleria* grasslands in Figure 7-2a.

Forest product collection was also an important component of livelihood strategies in Nakat and in the entire Baram area. Villagers from Nakat collected forest products mainly along the Bakong River, and also in other watersheds far from the village. In the swamp forests along the Bakong River, they collected forest products such as *jelutong* (*Dyera* spp.) and rattan which they sold to Chinese merchants. Chinese merchants had settled in Marudi in 1883 (Chew 1990: 74) and periodically visited the village to buy forest products. Later, the merchants from Kuala Nyabor took over this role, according to the Nakat villagers and the merchants of Beluru. According to them, Chinese settled and started commercial activities in Beluru around 1920, and merchants opened shops at Kuala Nyabor around 1930. The merchants played an important role in the Iban's livelihoods, as they bought forest products from the Iban and sold them everyday goods. The villagers also recall that the Pacific War (1941–1945) was a very difficult time for them, because all the Chinese of Bakong River were forced by the Japanese occupation forces to stop their commercial activities. Products such as sugar, salt, and clothes became unavailable with the suspension of the Chinese merchants' activities.

The proportion of time distributed between dry and wet rice cultivation and forest products collection changed periodically. As shown in *Sarawak Gazette* records, forest products collection was actively done in periods of poor harvests caused by drought, flood, and infestation of insects and wild animals (SG 17/1/1910: 26; 1/4/1915: 79). Conversely, collection was abandoned when prices of forest products were low, or during the season that rice growing demanded the most intensive use of labor, mainly the harvesting period (SG 1/4/1931: 87, 2/12/1935: 234). The area of rice fields was significantly reduced in the villages near Miri when many villagers started to work in the petroleum industry (SG 2/1/1930: 25, 1/8/1930: 203). In general, activities would shift depending on environmental and economic conditions, although forest product collection and rice growing continued to be an integrated part of a mixed portfolio of economic activities.

To summarize this period, the founding of Nakat village and the pioneering land use of its early inhabitants, was a results of Brooke's desire to maintain peace in this region close to Brunei and to encourage forest product collection. On the other hand, the Iban who moved to this region were driven by land and natural resource shortages they faced in the Third Division where they came from. Once they arrived in their new territory, the new settlers aggressively exploited the primary forests to produce a decent livelihood and to increase landholdings. The villagers adopted mixed resource-use strategies. They created a complex landscape, converting both hill and swampland forests to grow different kinds of crops. They also incorporated forest product collection in order to supplement their income when agricultural activities yielded low returns. Therefore, though the early settlers had responded to the government's land settlement schemes, they also created their own version of resource use, and were not just "obeying" what authorities had told them to do.

The Period of Commercialization of Rice and Rubber (1950–1985)

The years from 1950–1985 are characterized by greater commercialization of rice and rubber and with it villagers' changes of economic emphasis from dry rice production and forest products collection to wet rice production and rubber tapping. This period was influenced by three major historical events: the end of the Pacific War in 1945, Great Britain retaking control of the region in 1946, and Sarawak's entry into the Malaysian federation in 1963. Land use and economic activities in Nakat were directly and indirectly affected by those events. The end of this period is marked by a significant turning point of the impact of the Miri economy on the villagers' economic activities.

The years between 1950 and 1985 are marked by the growth of the commercial logging industry, which by the 1970s and 1980s had become thoroughly established in the whole of Sarawak. Although before the Pacific War (1941–1945), logging activities had already begun, their scale was small and the timber was mainly for the domestic market.[7] By contrast, after the war, timber was mainly exported abroad. The scale of logging grew following increased demand and developments in logging technology (Kaur 1998). In the Baram watershed ramin (*Gonystylus bancanus*) found in swamp forests became targeted for logging from the 1950s to early 1960s. The government recognized favorable conditions for commercial logging in hill forests in the early 1960s (SG 30/9/1963: 230). Since the late 1960s logging in dipterocarp forests in lowlands and hills became widespread (SG 31/7/1969: 180), while logging concessions were held mainly by Chinese entrepreneurs.

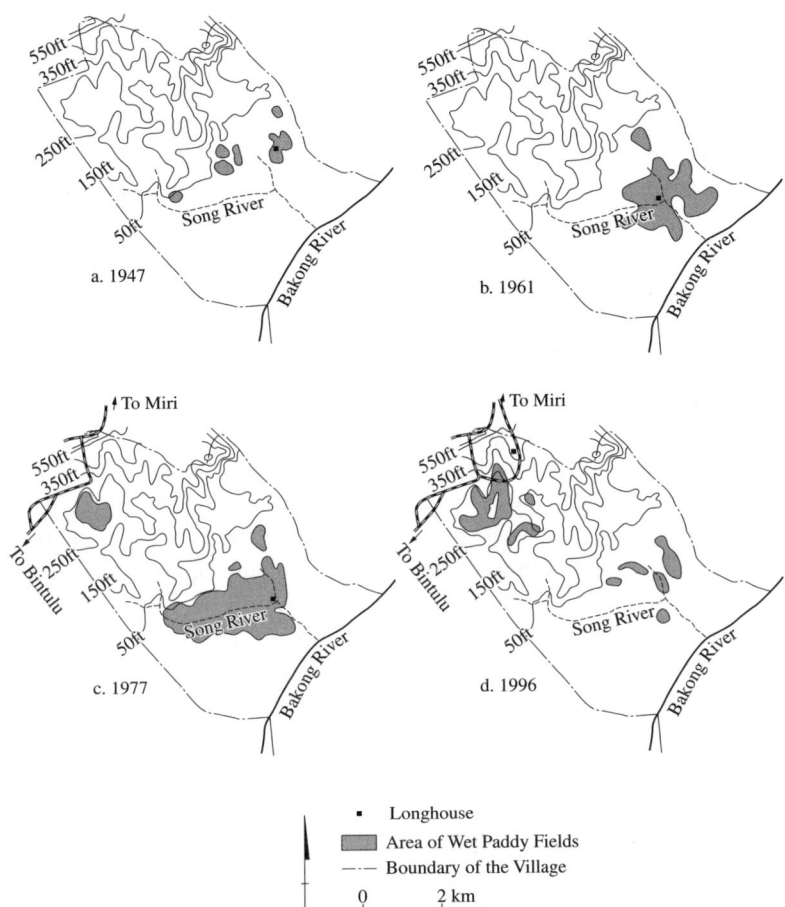

Figure 7-2. Changing land use of Nakat village, 1947–1997

According to the merchants of Marudi, before the 1960s a demand for rice grew in towns such as Miri and Marudi, which were recovering from the Pacific War. This rice was supplied by the villages along the Baram River. After the 1960s, a lot of logging camps started to appear, which needed a large quantity of rice. Furthermore, following the international food crisis, in the 1970s prices for rice increased across the board.[8] These three reasons boosted rice production from the study area. As a result, the 1960s and 1970s were marked by a dramatic commercialization of the rice production in the Baram area. According to the merchants of Marudi, the Bakong watershed in the Baram area was the most productive for rice production because of the availability of large swamps that could be converted for wet rice-growing. In addition, the local Iban had adequate experience with market-oriented production. According to the merchants interviewed, relative to other ethnic groups in Baram, the Iban have a stronger tendency to produce goods that have market value.

According to the villagers of Nakat, the number of those who worked in logging camps and sawmills in 1960s and 1970s grew. Road and house construction work also became major wage earning opportunities in the 1970s. It has often been argued that logging affected local people's lives negatively by, for example, degrading the natural environment. Furthermore, work in timber camps entails physical danger (see Hong 1987). Villagers from Nakat, however, did not see themselves only as victims of the logging industry as they took advantage of the opportunities that it brought and benefited economically. In a way, the people living in the regions where logging took place became essential to the logging operations, as they supplied labor to the timber operators and food to everybody in the camps. Because of this change in economic activities, the number of villagers who engaged in forest products collection declined after 1960.

Wet Rice Cultivation in Nakat

The growing of rice in swamplands produces higher yields per land unit than rice grown on dryland. Thus, after the pioneer period, the villagers rarely opened rice fields in the primary forests that remained on the periphery of the village, but turned to secondary forests and grasslands on swampland near the longhouse. Figure 7-2, derived from aerial photographs,[9] shows the distribution of lands converted to wet rice fields. Towards the end of the pioneering period, the distribution area was small (Figure 7-2a). In 1963, when rice became a highly sought after commodity, the area under wet rice cultivation grew (see Figure 7-2b). By the mid-1970s, when the production of rice was at its peak in Nakat, the distribution of rice fields had increased again and reached its maximum extent as shown in Figure 7-2c. Table 7-1 provides the areas of wet rice fields and numbers of households during two periods, and percentage increase of each of these variables.

The villagers claim that there were no drastic changes in the number of household members which suggests that household increase and village

Table 7-1. Wet rice production and village expansion in Nakat

YEAR	AREA OF WET RICE FIELDS[1] (HA)	NUMBER OF HOUSEHOLDS[2]	PERCENTAGE INCREASE OVER 14/16 YEAR PERIOD	
			WET RICE FIELD	HOUSEHOLDS
1947	75	29	-	-
1961	297	46	296.0	58.6
1977	534	55	79.8	19.6

Source: 1. Measured from Figure 7-2.
2. Ichikawa, field notes.

population growth were in proportion. This means that the growth in lands under cultivation far exceeded the rate of population increase (Table 7-1). The reason for this accelerated growth in rice production was the increase in the market value of rice and families responding to this change by expanding their rice area. Since 1960, some households were said to have increased the size of their rice fields to an estimated four to five ha per field.[10]

The rice fields became concentrated along the Song River, which is located rather far from the longhouse. Producing rice along the Song River had the advantage of facilitating the transportation of rice in boats. According to the villagers, carrying home harvested rice is a very strenuous part of rice farming, especially when fields are far from the longhouse. In the village area, the Song, which is three to four meters wide, is the only river that is navigable by longboat. Large amounts of harvested rice usually are transported by longboat from the rice fields to the longhouse or to Kuala Nyabor to be sold to Chinese merchants. The other reason for choosing the Song River is the existence of a large swamp along its banks.

In the first year of planting, farmers cleared fields in areas of secondary forests, while in subsequent years fields were opened in *Screlia* grassland. Single fields could be cultivated for three successive years. In the fourth year, when weed infestation became excessive, new patches of secondary forests would be cleared. This was the typical way of wet rice growing in that period. Also, the villagers explained that some farmers would open new fields not in forest but rather in dense *Screlia* grassland. Those fields were used for one or two years before shifting again. In fields opened in secondary forests and dense *Screlia* grassland, there are few obnoxious weeds because the shade of the secondary forest and the dense *Screlia* had suppressed these weeds. During this period, rice planting was always done by broadcasting seed, even though the transplanting method would have been better to reduce weed infestation. According to the villagers, the lack of labor, most of which was used to expand the rice production area, made it impossible to use transplanting technology. Weeding also consumed too much time and capital. The villagers controlled for weed problems by selecting lands shaded by fallow forests or dense grasses. By applying the broadcasting planting technique and a short rotation fallow system, they could expand and manage extensive wet rice field areas.

Increase of Para Rubber Plantation

Besides wet rice growing, rubber tapping was another important activity in the village from around 1950 to 1980. In the Second and Third Divisions of Sarawak, Para rubber trees had been actively planted since the early 1900s (Cramb 1988: 112). In the Bakong area, however, there were few planted rubber gardens, because the price of wild rubber was so good that local people were less motivated to plant Para rubber (SG 16/3/1917: 73). Rubber cultivation was also constrained by international regulations controlling market prices.[11] In Nakat village, rubber growing began during the war and was actively continued during the 1950s and early 1960s when the price of rubber had increased (Figure 7-3).

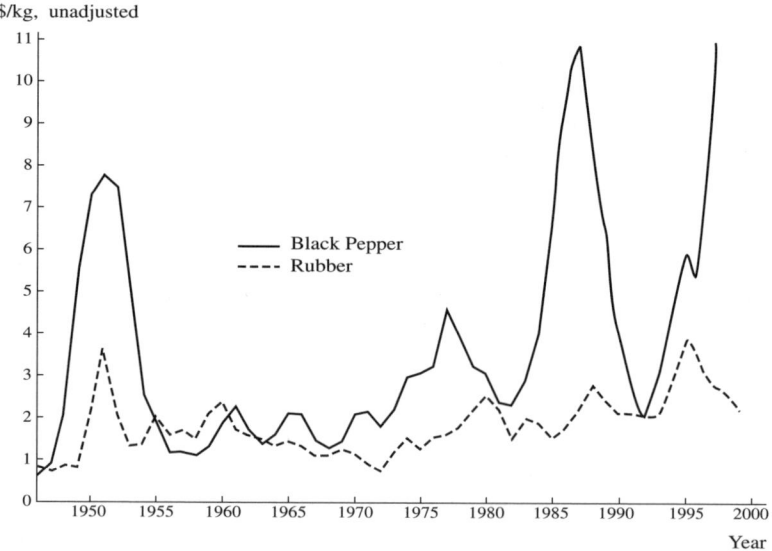

Figure 7-3. Change in value of black pepper and rubber

Rubber gardens were located on abandoned fields after hill rice had been harvested, so that the size of the each rubber plantation that time may have been around one ha. Since the hill rice fields were opened every year in various places around the village, the rubber gardens were also widely spread out, although they were more intensively located near the longhouse (Figure 7-2c). Rubber tapping was practiced in those gardens located near the wet rice field that was being cultivated during a particular four-year cycle. The proportion of time spent in rice growing and rubber tapping varied annually depending on the price of each commodity. When rubber prices were high, rice growing was cut back. Villagers dedicated time to rubber tapping activities even to the extent of refraining entirely from rice growing. Conversely, the value of rice as a marketable commodity rose in the latter 1960s and early 1970s when the price of rubber stagnated (Figure 7-3).

Another element encouraging the increase of these activities was simply the availability of labor. As already mentioned, unmarried young men would migrate to find work outside the village. This included wage work in sawmills, logging camps, and public construction. Most of the young men who worked outside, however, in this period usually returned to the village when they got married. They would then remain in the village, dedicating themselves to rice cultivation and rubber tapping, the most profitable activities at the time. In Iban communities mainly males do the felling of forests to make rice fields. As many male laborers remained in the village in the 1960s and 1970s, the task was smoothly accomplished. Under these conditions, forests in the village were aggressively converted to rice fields and rubber gardens.

The Period of Commodity Diversification

Until the 1980s and first part of the 1990s, the logging, petroleum, and oil palm industries around Miri were prospering progressively. In Nakat at the same time, a number of products from the village, such as rattan baskets, edible forest products, river fish, pepper, fruits, fowl, and pigs, became marketable items. The reasons for this diversification and commodification of what used to be household consumption items are explained below.

The Miri-Bintulu road, constructed from Miri, passed through the northern part of the village in 1965 and connected it to the Bintulu road in 1974. Road construction was one of the development projects started when Sarawak joined the Malaysian Federation in 1963. Before the advent of the road, it would take a day's walk on forest paths to reach Miri from Nakat. By road, the same destination was just an hour's drive by car, although the road was at that time still unpaved and not in the condition it is in today. After the road was built, however, most of the villagers continued to live in the longhouse along the river,[12] as it was a convenient place for rubber and wet rice production, and close to the Chinese merchants of Kuala Nyabor. To get to Miri, Nakat villagers would first go to the Miri-Bintulu road on foot (a walking distance of about two hours) and then take a bus. However, during the 1970s and early 1980s, signs of impact from the road construction and Miri economy could be observed on the land use patterns of Nakat. In the early 1970s some 25 households opened pepper gardens with subsidies from the Agricultural Department, on the way from the riverside longhouse to the road (Figure 7-2c). The harvest from these gardens was sold to merchants in Miri, who paid better prices than in Kuala Nyabor. Figure 7-2d also shows a wide area of bush fallows with only a few years of abandonment. Unfelled forests distributed next to the road was cleared by many villagers to claim usufruct rights. Villagers now also gained profits from selling vegetables grown in their swiddens, which they sold in huts built along the roadside.

Recovering from the disasters of war, Miri grew rapidly after 1960s as it prospered from the logging and petroleum industries. Commercial logging in the Baram River area continued until the early 1990s and petroleum production was also increasing.[13] In the area between Miri and Bintulu, the oil palm industry rapidly developed after the construction of the Miri-Bintulu road.[14] As these economic changes took place in the region around Miri, its urban area expanded rapidly, especially in the 1980s and early 1990s. The population of Miri increased more than twofold from 52,000 in 1980 to 130,000 in 1991 (Department of Statistics, Malaysia 1997). After the mid-1980s, access from the village to Miri improved as a result of the improvement and widening of roads, and the public transportation system.[15] Under these conditions, longhouse communities in the Bakong area who had easy access to the road began to choose Miri as the place to buy daily necessities and sell new products. As a result, the amount of trade with Kuala Nyabor merchants declined. By 1985, all the four Chinese shops there had abandoned their operations and moved to bigger cities or towns such as Miri and Marudi.[16]

The withdrawal of the Kuala Nyabor merchants, who had been the primary buyers of rice and rubber, made the value of these commodities drop drastically in Nakat. If the villagers wanted to sell them, they had to go up to Beluru, which was a one-hour journey by motored longboat; it was costly in terms of fuel as well as time. The villagers also lost their usual place for buying daily necessities such as sugar, salt, coffee, and canned foods.[17] However, the villagers swiftly adjusted. After 1985, various kinds of products other than vegetables grown in swidden fields were also sold along the Miri-Bintulu road, such as rattan baskets,[18] rattan and fern shoots, and fruits. Villagers also turned to selling these products on the streets of Miri. Pepper and fowl were sold to Chinese merchants in Miri.

After the Kuala Nyabor Chinese had moved out in 1985, the villagers became totally dependent on Miri as the place to buy goods and sell commodities. The small huts built along the Miri-Bintulu road to sell products, also served as temporary homes en route to Miri. The original longhouse increasingly only became occupied when busy with agricultural work. Finally, in 1995, villagers started to build a new longhouse along the Miri-Bintulu road. Their living base had completely shifted from the riverside to the roadside.

The villagers themselves say that before the road existed, they sold only rice and rubber. Subsequently, they became aware of other potential commodities and thus diversified their economic strategies.[19] After 1985, although a few villagers sold only rice and rubber,[20] almost all the villagers obtained income from other activities. The villagers considered rice and rubber production as hard work, while giving lower returns than the same efforts in other income generating activities. Timber from primary forests and mature secondary-forests in Nakat was logged and sold to small Chinese entrepreneurs from the mid 1970s until the 1990s. Wage work in and around the rapidly growing city became especially important in sustaining household incomes. Today, young and senior villagers work for wages even after they are married, and almost all the unmarried women (who formerly always remained in the longhouse) work outside the village.

A slight reversal of this trend, however, occurred when rice became increasingly sold again since 1995. Nakat villagers started selling rice on the streets of Miri to city dwellers who reckoned that local rice was of better quality than the imported varieties. Compared to rice sales in the 1970s, however, both the number of households engaged in selling rice (just six families in 1995) and the quantity sold were much smaller. Following the shift of longhouse in 1995, many rice fields were opened again near the new longhouse. Since then the number of rice fields next to the river declined. The wet rice was in most cases grown in small fields, which were opened in grasslands using herbicide and/or bush knives. Although planting was partly carried out by broadcasting, transplanting became the main planting method, which the villagers felt was a more reliable and productive method. Table 7-2 shows the changes in location of rice fields between 1995 and 1997.

The number and type of economic activities the villagers engaged in varied per household. For example, in young households, the main income came from the husband's cash-earning work. The wife was responsible for childcare and sometimes worked small rice fields herself. In households that had more than two generations of workers (such as a senior and a young couple), the younger man would work outside the village, while his wife and the senior couple would engage in a number of activities, including wet and dry rice growing, collection of rattan, making of rattan baskets, pepper cultivation, fishing, and livestock raising. The critical constraint in any household's selection of economic strategies was the number of working persons.

Government policies also affected people's economic activities. According to the villagers, government services such as agricultural subsidies became available since increased communication with Miri. After the late 1980s, when the price of pepper was high (Figure 7-3), almost all the families prepared pepper gardens, and all of them had received subsidies from the Agriculture Department to do so.[21] The subsidies that villagers have received recently could be used for planting high yielding rubber and fruit trees, raising fowl and fish, and buying fertilizers to grow rice. The villagers said that after they moved to the roadside, communication with the extension officers became easier and it was easier to apply for subsidies at government offices in Miri.

Given these spatial and economic changes, there have also been effects on the landscape. Natural vegetation has grown back in the lands around the former longhouse by the river, formerly cultivated with rice and rubber. The regeneration process, in other words, has begun. On the other hand, along the road, after the relocation of the longhouse and agricultural fields,

Table 7-2. Number of rice fields by location, Nakat, 1995 and 1997

Year	Number of river side rice fields	Number of road side rice fields	Ratio of road-side rice fields to traditional river side
1995	30	20	0.67 : 1
1996	13	32	2.46 : 1
1997	9	41	4.56 : 1

Source: Ichikawa (2000b)

the landscape has been transformed from forests to pepper and fruit gardens, home gardens, and many rice fields.

Conclusions

As outlined here, land uses in the village of Nakat can be distinguished in three different periods, over a total 100 years. Each of these periods relates to a set of political, social and economic conditions outside the village.

The beginning of the village itself in the Baram area was a result of wider political and economic factors: the acquisition of new territory by Brooke and the encouragement to the Iban to move to this region, as well as official interest in forest products collecting. For the Iban on the southwestern part of Sarawak, official actions proved to be a stimulus, as they themselves were beginning to face problems of land and resource shortage. The Baram area was still abundant in land and forest. After the settlement, the villagers aggressively converted primary forests to rice fields and collected forest products. When rubber prices in the international market went up after the Pacific War, there was an expansion in rubber gardens. Also in the same period, following the growth of demand for rice as logging camps and new arrivals appeared in sizable numbers in the Baram River at the end of the 1960s and 1970s, the villagers increased their rice production. A large area of rice fields was formed along the Song River, which was not only suitable for extensive new tracts of rice fields but for transporting the larger amounts of harvested rice. Secondary forests were used for rice growing, once they had formed after initial rounds of rice production. The wet rice fields were shifted every four years or so.

In the 1960s and 1970s, a large area of forest in the village was converted to rubber gardens and wet rice fields. From around 1985, various kinds of products from the village started to be traded as a result of the overall development of the region, specifically road construction, expanding urban centers, and favorable government policies for agriculture. In that period, various kinds of forest products became marketable commodities and started to be sold along the Miri-Bintulu road and in Miri. Since the villagers started to live along the road, the forests were converted to pepper gardens, home gardens and rice fields.

In the three periods land use changed profoundly in Nakat. However, at the same time there were important fluctuations within a period. Economic emphases could change depending on what the wider conditions were at any point in time. In the pioneer period, for example, villagers would choose either rice growing or forest products collection as their main activity. In the period of commercialization of rice and rubber, they would shift emphasis between rice growing and rubber tapping. Presently, they select among a range of activities including pepper cultivation and weaving rattan baskets.

Dynamics could go as far as modifications within one single economic activity. For example, when the price of rice was high in the 1960s and 1970s, large rice fields were made on forest fallows, using broadcasting as the dominant planting method. Now that the relative importance of rice has declined, small fields are made in grass fallows while the dominant planting method has become transplanting. The villagers have adapted to the changing economic conditions by adopting different agricultural methods in their land use.

While these changes took place in the village, no serious land degradation did occur, such as *Imperata* grassland formation. The Song area of the former large wet rice fields has evolved into bush and forests. Pepper production in vicinity to the Miri-Bintulu road started roughly 15 years ago in the village. Since the productivity of these gardens is now dropping, the pepper gardens are being abandoned. During the fieldwork, such abandoned pepper gardens observed along the Miri-Bintulu road reverted to bush that seems to be able to recover into forest in the future.

Adaptations to outside conditions seemed to be part of the villagers' strategies for surviving in a severe and unstable natural environment. As already mentioned, the villagers have experienced crop losses due to natural disasters, including drought, flood, and pest infestation. In order to counter-balance such risks, it may be necessary for them to interact increasingly with the outside economy. The villagers seem to have chosen the more beneficial activities and appropriate methods of land use through careful consideration of conditions outside the village. Nakat inhabitants depend not only on their village economy, but adjust their livelihood strategies to respond positively to political, social and economic situations developing outside of the village. The degree to which they depend on one versus the other has changed throughout village history.

This paper is focused on a single village. Similar changes of land use and economic activities, however, may be observed in other areas of Sarawak as well as in other parts of Southeast Asia, where population density is low and tropical rain forests are dominant. Cramb (1988) noted that drastic changes in the Iban's activities and land use have occurred as a result of wider economic conditions in Sarawak. Padoch (1982) reported the change of longhouse locations and today many such cases can be observed along the Miri-Bintulu road. There are similar reports for South Sumatra (Takaya 1979; Furukawa 1994) and colonial Malaya (Kato 1991). The livelihood strategies recognized in Nakat are by no means unique.

On the streets of Miri, the number of Iban and other ethnic groups from local villages selling various local products are said to be increasing recently. Their land-use practices are likely to change with such changes in their economic strategy. Today in the Baram area, the road network is rapidly expanding further upstream as a result of commercial logging. Before, in the interior, river transportation was the only way of communication but today the former logging roads play an important role in improving communication and exchange between an ever larger number of places. The people from upstream villages now can go to Miri faster and cheaper. Following such changes, land uses and resource exploitation of upriver people, of the Kayan, Kenyah and others, may follow similar patterns of change presented in this paper.

Acknowledgments

Research was supported by grants from The Osaka International House Foundation (1996) and The Daiwa Bank Foundation for Asia and Oceania (1998).

Notes

1 The land use types observed were primary forest, secondary forest, bush, rice field, rubber garden, pepper garden, and so on. Among these, information on wet paddy field is based on Ichikawa (2000b).
2 In this study, the person recognized as the member of family (*se-bilek*) by the chief of household (*tuai bilek*) was counted as a "villager." Out- and in-flow of villagers for wage work outside the village was always observed. The number of villagers who stayed for more than 6 months from June 1995 until May 1996 was 262. The rest of the villagers stayed outside the village for more than 6 months or even for years.
3 Sarawak was ruled by James Brooke, an English adventurer, and his descendents, Charles Brooke and Vyner Brooke, from 1841 to 1941.

4 Sandin (1994: 227–228) describes in detail the Iban's move to the Baram River.
5 Even before the settlement started, many Iban, who were famous for *bejalai* or traveling, went to Baram to seek forest products (SG 1/3/1883: 26, 1/5/1884: 44). These travels might have led them to discover the abundance of forest products in the area.
6 'Secondary crops' here refers to those crops grown in the space or interval between the rice plants, such as a variety of vegetables and root crops.
7 For example, in Baram River a sawmill was established around 1920 to produce lumber for construction purposes in Miri (SG 2/8/1920: 176).
8 The price of rice in Sarawak, which was controlled to some extent by the government, had increased 1.4 times from 24.46 ringgit per 100 kg in 1971 to 33.90 ringgit per 100 kg in 1974 (Department of Agriculture, Sarawak 1981).
9 The wet rice fields, especially those that are cultivated successively for a few years can be seen even from a distance as grassland. Figure 7.2 was made by tracing the grasslands and bush on the aerial photographs in each year. According to the villagers, if a rice field is abandoned, the vegetation in the area will grow from grass to bush within three to seven years. Therefore, though the figure does not necessarily show the exact rice fields cultivated in the year when the photograph was taken, it shows the approximate area of rice fields in that period.
10 The area is estimated from recall interviews about the amount of rice produced at that time. According to the villagers, families who had larger rice fields opened around ten acres (about four ha) and in favorable years around 6,000 kg of rice was harvested. Assuming productivity to be 1,200 kg/ha (cf. Ichikawa 2000a), the average size of a family's rice field was roughly five ha.
In the study year in 1995–1996, 38 households (45 per cent of the total number of household) cultivated wet rice, for which the largest field was 1.48 ha and the average was 0.54 ha.
11 In order to control the market price of rubber, the International Rubber Regulation and the International Rubber Agreement were concluded in 1925 and 1934 respectively among the rubber producing countries (Barlow 1978:62). In Sarawak, the planting of rubber trees and the amount of latex sold were controlled (Cramb 1988:115–116).

12 According to the villagers, although eight families moved to the roadside in 1964 when the road was still under construction, 47 families remained in the original longhouse.
13 The amount of exported logs had increased up to 15 million m³ by the early 1990s, and exported petroleum also increased to nearly 10 million tons in 1996 (Department of Statistics, Malaysia 1997).
14 The plantation areas of Sarawak have drastically increased especially in the 1990s and were 160,000 ha in 1996 (Department of Statistics, Malaysia 1997). The center of plantation development has been the area between Miri and Bintulu (King 1993: 277).
15 For example, bus services started in the early 1970s and in the 1990s the buses from Miri to the village departed almost every 30 minutes from 6 in the morning until 4 in the afternoon.
16 According to interviews with the villagers and Chinese merchants of Burulu.
17 Although foodstuffs could be bought in Miri, they had to travel on foot for two hours between the Miri-Bintulu road and the longhouse.
18 Rattan baskets are not a traditional product, but they were made for selling by the villagers since around 1985. Initially, some villagers made them by imitating existing articles made and sold by Chinese.
19 The villagers said that there were no government programs and directions to encourage them to produce commodities. They started selling their commodities in Miri after the mid-1980s, following the local Iban who had previously sold goods in Miri.
20 In 1995, rice growing engaged less than half of all the households and the rice was sold by six households, while rubber tapping was practiced by almost none of the households.
21 Subsidies took the form of pepper buds, fertilizers, weed killers, and some cash for preparation of posts for pepper planting.

Bibliography

Barlow, C. 1978. *Natural Rubber Industry*. Kuala Lumpur: Oxford University Press.
Chew, D. 1990. *Chinese Pioneers on the Sarawak Frontier 1841–1941*. Singapore: Oxford University Press.
Cramb, R.A. 1988. The Commercialization of Iban Agriculture. In R.A.

Cramb and R.H.W. Reece, eds., *Development in Sarawak*, 105–134. Clayton: Center of Southeast Asian Studies, Monash University.

Department of Agriculture, Sarawak. 1981. *A Digest of Agricultural Statistics*. Kuching.

——— 1991. *Agricultural Statistics of Sarawak 1990*. Kuching.

Department of Statistics, Malaysia (Sarawak Branch). 1997. *Yearbook of Statistics 1997 Sarawak*. Kuching.

Dove, M.R. 1985. *Swidden Agriculture in Indonesia*. New York: Mouton.

Freeman, J.D. 1955. *Iban Agriculture: A Report on the Shifting Cultivation of Hill Rice by the Iban of Sarawak*. London: H.M.S.O.

Furukawa, H. 1994. *Coastal Wetlands of Indonesia*. Kyoto: Kyoto University Press.

Grossman, L.S. 1981. The Cultural Ecology of Economic Development. *Annals of the Association of American Geographers* 71(2): 220–236.

Hong, E. 1987. *Natives of Sarawak*. Pulau Pinang: Institut Masyarakat.

Ichikawa, M. 2000a. Sarawak-shu Iban sonraku ni okeru sicchiden inasaku: Uetsuke houhou ni miru tekiou senryaku (Swamp Rice Cultivation in an Iban Village of Sarawak: Planting Method as an Adaptation Strategy). *Southeast Asian Studies* 38(1): 74–94 (In Japanese).

——— 2000b. Sarawak-shu Iban sonraku ni okeru idou sicchiden inasaku no hensen (Transformation of Shifting Swamp-Rice Cultivation in an Iban Village of Sarawak, Malaysia). *Southeast Asian Studies* 38(2): 226–248 (In Japanese).

Kato, T. 1991. When Rubber Came: The Negeri Sembilan Experience. *Southeast Asian Studies* 29(2): 109–157.

Kaur, A. 1998. A History of Forestry in Sarawak. *Modern Asian Studies* 32(1): 117–147.

King, V.T. 1993. *The People of Borneo*. Oxford: Blackwell Publishers.

Land & Survey Department, Sarawak (Miri branch). Aerial photographs taken in 1947, 1961 and 1977.

Padoch, C. 1982. *Migration and Its Alternatives among the Iban of Sarawak*. The Hague: Martinus Nijhoff.

Peluso, N.L. 1992. *Rich Forests, Poor People*. Berkeley: University of California Press.

Pringle, R. 1970. *Rajahs and Rebels: The Iban of Sarawak Under Brook Rule, 1841–1941*. Ithaca: Cornell University Press.

Sandin, B. 1994. Sources of Iban Traditional History. *The Sarawak Museum Journal* 67. Special Monograph No. 7.

Takaya, Y. 1979. Agricultural Landscape in the Komering River Basin,

South Sumatra. *Tonan Ajia Kenkyu* (*Southeast Asian Studies*) 17(3): 444–466 (In Japanese).

The Sarawak Gazette. 1/3/1883; 1/5/1884; 1/2/1902; 4/4/1907; 17/1/1910; 1/4/1915; 16/3/1917; 2/8/1920; 2/1/1930; 1/8/1930; 1/4/1931; 2/12/1935; 30/9/1963; 31/7/1969.

Watt, M. 1983. *Silent Violence: Food, Famine and Peasantry in Northern Nigeria*. Berkeley: University of California Press.

8
The Ecological-Economics of Non-Sustainable Development: Logging Tropical Forests in Southeast Asia and the Pacific

Herb Thompson

Logging, population expansion and transmigration are predominant forces driving deforestation in Southeast Asia and the Pacific. The logging of these forests has largely been a function of the aggregate demand of Japanese consumers (Nectoux and Kurida 1989: 32). To date, international trading and logging capital has simply maximized its return by acting as a conduit between the major consuming nations of Japan, Korea and Taiwan, and major producing nations of Indonesia, Malaysia, the Philippines, Papua New Guinea and the Solomon Islands. Agents of the producing nations have typically sold off timber too cheaply, sacrificing public revenues and the undervalued non-timber benefits of the standing forest.

There have been numerous studies of the effects of logging, in Southeast Asian rainforests, focusing on the amount and types of damage sustained by the residual stand immediately after logging, and the degree to which the forest floor was disturbed by roads and tracks (Borhan *et al.* 1990; Cannon *et al.* 1994: 59–64; FAO 1998; Sist *et al.* 1998). Most site impacts associated with logging operations in the tropics can be attributed to skidrails or to haul roads, and for the most part these are soil-related impacts. Some of the major consequences due to the destruction of tropical rainforests include loss of genetic resources; destruction of the process of biomass production; disruption/disappearance of the cultures of indigenous peoples dependent on the rainforest.

The general aim of this paper is to begin to understand, examine and present the regional forestry of Southeast Asia and the Pacific sector and its interdependencies. A congruent aim is to pursue an analysis that recognizes the congruity between ecological communities and the environment in terms of both political economy and political ecology. This is structured around both an explicit and implicit critique of the methodological practices of neoclassical economics. Because mainstream analysis dichotomizes humanity and nature in its analysis of rainforest logging, neoclassical economists are unable or unwilling to incorporate those areas of most concern, of which loss of biodiversity or decimation of indigenous communities, are but two examples.

In the section below the relations between the *modus operandi* of development and nature with respect to tropical deforestation is identified. This is broken down into two sub-categories of political economy and political ecology for exemplification. A theoretical synthesis is pursued in the section on the synecological relations of production and distribution. Some attempt is then made in the penultimate section to generate an optimistic perspective, noting that not all is doom and gloom; followed by concluding comments and a summary.

Development and Nature

The Political Economy of Development

Economic development should increase a country's material prosperity and social welfare, reduce poverty and inequalities, improve the quality of life of individual citizens by provision of better health, education and other services, and increase the quality and quantity of economic output by better access to technology. However slowly development occurs, it involves major changes throughout the social formation, many of which can impinge on forests (agriculture, roads, logging for timber, fuelwood, etc.).

Just as deforestation accompanied the economic development of many of today's wealthy countries, so it continues unabated as part of economic growth in the less developed countries. Forests have been an essential basis of human prosperity, providing diverse products and services throughout the evolution of humankind (Perlin 1989). The Earth Summit of 1992 epitomized the failure of Western governments and environmentalists to persuade representatives of tropical nations to accept international supervision of their rainforests. Representatives of Malaysia, in particular, resisted any suggestion that natural resources should be "internationalized." The only agreement reached was a general statement about balancing forest exploitation with conservation (*The New Scientist* June 20, 1992: 5).

Today, for most of the world's population the alleviation of poverty is primary, with economic development playing the leading role. Many would argue that less developed countries have little if any responsibility for global negative externalities created by tropical deforestation. The pressures that exist on less developed countries not to clear the tropical rainforest is at least partly due to the unwillingness of those in rich countries to control their consumption of fossil fuels. "The idea that the third world must share the burden for present problems such as global warming is an example of environmental colonialism"(Agarwal and Narain 1991: 1). From the perspective of policy-makers in less developed countries, in order for alternative forest uses to be a viable option, they must yield net returns to developing countries that are greater than those derived from timber management (Clark 1976; Ozanne and Smith 1993). According to Smith *et.al.* (1995: 31), "moderate" environmental non-government organizations understand that "economic incentives are necessary to encourage using tropical lands to produce wood."

Simultaneously, no human activity, aiming at economic development can succeed without the normal functioning of the ecosystem. In that sense, the integrity of the ecosystem must be regarded as the foundation for economic and social progress. Human productive capacity requires sustaining the ecological balance (Wang and Xu 1996: 2); yet only recently has ecological protection emerged as a significant theme and policy goal pursued by development economists (Batabyal 1995; Chichilnisky 1994).

After World War II a market developed, and quickly grew, for low-density rainforest timbers to be used for veneer, plywood and general construction. Minor forest products, which had traditionally been predominant in international trade (eg., latex and rotan) were eclipsed by timber. Within the Southeast Asia/Pacific region, the main source of supply has moved successively. First was the Philippines where soon after the war North American logging companies brought their knowledge and equipment. Indonesia was next in the early 1970s when the trees were depleted in the Philippines, replaced in turn by Sarawak and Sabah; and most recently Papua New Guinea and the Solomon Islands have been added to the roster (Goodland *et al.* 1990; Whitmore 1993: 117; Thompson and Kennedy 1995). In 1992 when plywood prices were increased on exports from Indonesia, Sarawak introduced a temporary ban on cutting and export of logs. Then the price of the North American Douglas Fir jumped by 60 per cent, given a ban on log exports in Washington State. At this point Japanese importers turned their attention to Papua New Guinea and the Solomons. Imports of logs from Papua New

Table 8-1. Land and forest area of small island states and dependent territories

Region	Land area	Total forest 1995[a]	% of land area	Total forest 1990[a]	Change 1990–1995[a]	Annual change[a]	Annual % change
Africa[b]	1251	218	17.4	190	28	6	2.8
Asia[c]	93	4	4.3	4	0	0	0
Oceania[d]	8767	4964	56.3	5054	-90	-18	-0.4

Source: FAO (1999: 29)
a times 1000 ha.
b Africa includes: Cape Verde, Comoros, Mauritius, Réunion, Sao Tome and Principe, Saint Helena, Seychelles.
c Asia includes: Macau, Maldives, Singapore.
d Oceania includes: American Samoa, Cook Islands, Federated States of Micronesia, Fiji, French Polynesia, Guam, Kiribati, Marshall Islands, Nauru, New Caledonia, Niue, Northern Mariana Islands, Palau, Samoa, Solomon Islands, Tuvalu, Tonga, Vanuatu.

Guinea surged 65 per cent and the supply from the Solomon Islands grew 42 per cent (*Far Eastern Economic Review* March 4, 1993: 44).

A Global Conference on the Sustainable Development of Small Island Developing States was held in Barbados in April, 1994 (see Table 8-1). The important states of the Pacific, Papua New Guinea, Solomon Islands and Vanuatu, were major participants in this conference. These particular nations are important in this context because Japan is a major consumer of their timber products and logging companies from South Korea and Malaysia are some of the major players in the region. Because of this, the Pacific Island states cannot be analytically separated from Southeast Asia. They have been experiencing significant forest degradation due to heavy exploitation of timber resources (FAO 1997a). The people of these nations also have a high degree of household dependence on forests for a variety of wood and non-wood products.[1]

As corporate executives, workers, farmers and peasants have come to realize, long-term trends have shown falling prices for agriculture and mining products. By contrast, timber has maintained its real value as a commodity that makes the forest sector particularly important to less developed countries (Dargavel and Tucker 1992: 1). During the last half of the 20th century, trade balances for forest products show that countries in Northeast Asia increasingly imported raw wood materials (saw and veneer logs, pulpwood and woodchips) (Barbier *et.al.* 1993: 26–27).

Toward the end of the 1950s, Japanese demand—for hardwood logs as raw material for its timber-working industry—began to soar. It continued to rise rapidly through the 1960s. Between 1970 and 1985 Japan alone took half the world's rising import of sawlogs, most of it from Southeast Asia. For the first time, there was substantial demand for the dipterocarp timbers of Sundaland. At first, emphasis was placed on the group of *Shorea* species known as *meranti*, abundant in the lowland forests; but later, overall demand widened significantly (Brookfield *et.al.* 1995: 62–63). The loggers in tropical countries of Southeast Asia and the Pacific provided for this demand (Thompson and Kennedy 1996).

The required output also required new technologies. The one-man chainsaw first developed in the 1950s. This made possible a great increase in productivity coupled with transport improvements and much deeper incursion into the forest. The outboard motor, extensively used on the rivers and bays, bigger and more powerful trucks, four-wheel drive vehicles, the bulldozer, the crawler tractor for hauling logs and all the equipment used in modern highlead winching, brought with it a need for higher and faster returns. In the absence of any real knowledge of the resource and its ecology, and with totally inadequate means of enforcing such regulations as were already in place, concessions were offered by Southeast Asian (and now by Pacific) governments on extremely favorable terms. These were to be worked without supervision in order to develop log exports as important earners of foreign exchange (Brookfield *et al.* 1995: 63).

During the later 1980s, the forest sectors of Indonesia, Malaysia (Sabah), and the Philippines were making notable payments to their domestic governments of 400 million, 425 million, and 50 million USD a year respectively. Plywood and other panel products replaced log exports in Indonesia and the Philippines, the former for value-adding purposes, and the latter because the nation was basically running out of trees. Malaysia is now the largest exporter of unprocessed tropical logs, with lesser but significant quantities coming from Papua New Guinea and the Solomon Islands, also largely due to the investment of Malaysian companies. In Sabah, forest income accounted for 70 per cent of total government revenue (World Bank 1991: 72). These large timber revenue components of government budgets and export revenues also exist for Papua New Guinea and the Solomons. Poor people, through encroachment, are responsible for a significant share of the annual loss of forests, in part because they have higher discount rates than the nation's people as a whole because of the urgency of their current needs. On average, all participants in the forest onslaught may undervalue the longer-term benefits of investments in forests.

Alf Leslie (1987) has exposed the inadequacy of financial accounting, with its emphasis on short-term cash flow, to fully analyze the costs and benefits of alternative forestry investments. Valuation of nontimber tropical-forest products (food, construction materials, medicinal plants, fodder, and firewood) is plagued by inadequate measurement of costs, quantities extracted, and prices. The results of different studies cannot be directly compared because different methods have been used. Most such valuations seem to have been an afterthought (Godoy and Lubowski 1992: 424–430; FAO 1989; Liu 1996; Panayotou and Ashton 1992). The key gaps in traditional financial analysis are: failure to consider non-revenue benefits and external effects; market imperfections and distortions in the actual prices of inputs and outputs; and an arbitrarily determined, overvalued discount rate. In most less developed countries more trees are being cut than is desirable for national economic welfare. But even if economic sustainability can be achieved, the global net loss of trees, particularly in the tropical moist forests, will still be excessive from a global, rather than a national, perspective. Lele (1991) argues at a general level that: "[Sustainable development] is a 'metafix' that will unite everybody from the profit-minded industrialist and risk-minimizing subsistence farmer to the equity-seeking social worker, the pollution-concerned or wildlife-loving First Worlder, the growth-maximizing policy-maker, the goal-oriented bureaucrat and, therefore, the vote-counting politician." At least two of the five main forest services, biodiversity and regulation of climatic patterns, provide for global externalities and thus require preservation of forests, especially tropical moist forests.

Natural scientists frequently seem baffled or dismayed when mainstream economists display little appreciation of the gravity of environmental loss (Page 1995: 141). Many economists are convinced that the environment is a luxury good, that economic growth is essential, and only through economic growth will the environment be sustained (Beckerman 1992). Repetto sees it differently. He argues that it is the environment, whether or not its changes are measured, that will influence, and perhaps shape, the future development of the world economy (Repetto *et al.* 1989; Repetto 1992: 3–4).

The policy dichotomy between the economy and nature leads policy-makers to ignore or destroy the latter in the name of the former (Leiss 1974: 135–136; World Bank 1991: 34). One clear example, again unmeasured, is the fact that deforestation exposes us to mosquito-borne, tick-borne or rodent-carried diseases and the changing habitats of these disease-carrying insects (Lewontin and Levins 1996: 105). Logging creates the conditions that help to spread malaria, which is spread by a large variety of *Anopheles* mosquitoes. The situation is likely to spread uncontrolled because of the pools of water created on eroded land by logging and road construction (*The New Scientist* October 10, 1992: 5). Given the obviousness of the case, Arrow *et.al.* (1995) and Page (1995) have argued for the interdependency of the economy and ecology. According to these authors: (1) the environment is a collection of interlocking and nested systems as is the economy; the two sets of systems are themselves interlocking with the economy nested inside the environment; (2) equilibria that maintain these life support systems are sometimes fragile and can be undermined by economic activities; and (3) institutions and institutional incentive structures shape economic behavior.

Economists cannot value what the environment is worth; merely its value in market criteria. This means attaching cardinal valuations through monetary measures, such as prices and taxes, or through shadow prices by using contingent valuation, when ordinal valuations (more/less valuable) may be more appropriate, or useful. Conceptually, the total economic value (TEV) of a resource consists of its use value (UV) and nonuse value (NUV). Use values may be divided further into the direct use value (DUV), the indirect use value (IUV), and the option value (OV) or potential use value. Nonuse values include both existence values (EV) and bequest value (BV) (Table 8-2). Therefore, we may write:

TEV = UV + NUV, or
TEV = (DUV + IUV + OV) + (EV + BV)

Table 8-2 Disaggregation of total economic value

Total Economic Value				
Use Values			Non-use Values	
Direct use value	Indirect use value	Option value	Bequest value	Existence value
Output that can be consumed directly	Functional benefits	Future direct and indirect use values	Value of leaving use and non-use values for offspring	Value from knowledge of continued existence
Food Biomass Recreation	Ecological functions Flood control	Biodiversity Conserved habitats	Habitats Irreversible changes	Habitats Endangered species

Source: Barbier 1993: 216; Munasinghe and Shearer 1995: 35–36

Monetary valuations on their own capture the worth of the environment to particular groups of people, normally those engaged in market activity. We are given the "market" as metaphor, which seemingly accomplishes things by generating purposive behavior of self-interested, all-knowing, maximizing, individual molecules, of which it is constituted. Those who labor outside of the monetary market are omitted from the equation. For instance, the value women attach to the environment is usually invisible because the use they make of it is not, in many instances, subject to market valuation. The activities such as collecting firewood, gathering plants and fetching water, for both use and exchange, are vital for the sustainability of poor rural households but are seldom calculated in economic accounting terms. Similarly, the values we place on nature largely reflect our monetary priorities, not the use value of nature itself. Nature is constructed by human activity, and in seeking monetary values for environmental goods and services, nature is being transformed into a commodity (Redclift 1993: 14–15).

This transformation of nature into a commodity is particularly exemplified by those tropical nations in which international debt and trade imbalances exist in crisis proportions. There are many countries in which debt is a significant fraction of Gross Domestic Product, of which Malaysia, Indonesia, and the Solomon Islands stand out for particular attention (Kahn and McDonald 1995: 111). In order to meet current needs and overcome constraints, debt is normally acquired by less developed countries. Although the impacts of macroeconomic policies are difficult to analyze, and causal relationships even more difficult to infer, a high correlation has been indicated between exchange rate devaluation and tropical deforestation (Barbier *et al.* 1993: 54–55); as well as a positive relationship between tropical deforestation and public external debt (Kahn and McDonald 1990). Thus, the indirect impact of debt on increased tropical timber extraction and deforestation, through policy responses to correct or service the debt problem as implemented through structural adjustment programs, may be significant.

The Political Ecology of Nature
The richest of all the world's forest regions floristically is the Dipterocarp-dominated tropical rainforest found in the western part of the Indonesia-Malaysia area.[2] Eastern Indonesia, and points further east, are part of a contiguous rainforest area less dominated by the tree species *Dipterocarpaceae*. Malesia is a term used to refer geographically to this entire complex (Jacobs 1988: 1–6). The phytogeographic region referred

to as Malesia lies north and south of the equatorial line for over one-fifth of the world's circumference. Ten distinct subregions are normally included: Sumatra, Peninsular Malaysia, Java, Lesser Sunda Islands, Borneo, the Philippines, Celebes, the Moluccas, New Guinea and the Solomon Islands. These subregions can be re-divided into two major centers of tropical rainforest growth. The first large center of forest is in western Malesia, which consists of: Borneo, especially west of the central watershed, Peninsular Malaysia, Sumatra, the Philippines and Java. During the Ice Ages when the sea level was low and the shallow parts of the China and Java Seas were above water, this whole area, called the Sunda Shelf, or Sundaland, formed a continuous landmass. The forests of Sundaland are characterized by an enormously rich tropical flora for which volcanism, greater aridity and humans are responsible. Java, with 4500 species of native flowering plants and well represented by world standards, is no where near as rich as some parts of the rest of the area.

The second center of rainforest is on the island of New Guinea, with extensions as far as the Solomon's and Northeast Australia. Interestingly, the flora here is distinctly Asian in character even though the mammalian fauna is clearly of Australian derivation. After the break-up of Gondwanaland, New Guinea was pushed northwards in front of Australia into the tropics. By way of the geologically unstable area northwest of New Guinea, plants were able to migrate from Sundaland eastwards to New Guinea. There, a relatively poor temperate flora, which had for the most part succumbed in the heat of the tropics, was quickly superseded by the Malesian taxa. This was accompanied by a strong speciation, giving New Guinea its special character (Jacobs 1988: 125-126).

The two significant demarcation lines within Malesia are the "Wallace Line" and the "Torres Strait Line." The "Wallace Line" separates western from eastern Indonesia biologically (Wallace 1869). Monbiot (1989: 22-46) provides a first-hand description how, as one moves along the islands from Sulawesi to Irian Jaya, there is a dramatic change in animal life forms. The other demarcation line is the geographical boundary in the Torres Strait between New Guinea and Australia separating distinct floral features. Plant geographers argue that Papua New Guinea belongs to the Indo-Malesian floral region and zoogeographers place Papua New Guinea in the Australian region based on the distribution of animals and birds. This biological, zoological, geographical, political and economic complexity constitutes our region of concern.

Economies have dramatically altered the surface of the planet and are now influencing global biological, chemical, geochemical, and physical processes (Gowdy and McDaniel 1995; Turner *et al.* 1990). The rise of atmospheric CO_2 concentration leads many to predict that major global climate change is imminent (Tucker: 1995: 216). Human-synthesised chlorofluorocarbons are providing breakdown products that are catalyzing the destruction of stratospheric ozone with the consequent increase in life-threatening UV-B radiation at the Earth's surface (Kerr and McElroy 1993). Fisheries are in decline and soil erosion, desertification, salinization, and waterlogging are decreasing soil fertility in many places throughout the world. Alteration and simplification, on a world scale, of ecosystems is rampant (Houghton 1990).

Climate

Global climate change is one of the most complex science-derived issues to wind up at the center of political discourse. Therefore, our attention to the topic in this context will be limited to the context of deforestation. The theory of anthropogenic climate change is not a single, mathematical theory, but rather a cluster of interlocking theories, more or less defined. In considering the climatological implications of rainforest destruction, there are two major issues: the disturbance of the heat and hydrological cycles, and the impact of changed greenhouse gas con-centrations (House of Commons 1991: xix). We have a situation in which the time needed to give a reliable answer to the question of how will the climate change if large areas of land are deforested, is too long for the answer to be settled decidedly to the satisfaction of the present generation (Clark 1992: 39). The difficulty lies in quantifying the components of the energy and water balances in the undisturbed and disturbed ecosystems (Salati and Nobre 1991: 177). Logging and other forest operations do not, of necessity, have adverse hydrological effects (Bruijnzeel 1992), but most often timber harvesters and landless poor work in concert to deforest. The loggers provide access roads and remove the big trees, while the poor destroy the remaining vegetation and begin subsistence cultivation.

Traditionally, the causal relationship between vegetation and climate always viewed vegetation as the dependent variable, meaning of course that a large-scale change of vegetation would not by itself alter the climate. Recent empirical evidence appears to contradict this view, providing serious consequences for future generations. The studies are nearly unanimous in their findings: deforestation will lead to reduced rainfall and evapo-transpiration but increased air temperature (See Clark 1992; Shukla *et al.* 1990; Lean and Warrilow 1989; Henderson-Sellers and Gornitz 1984).

Both field work and computer simulation show that the consequences of deforestation will also mean that drought conditions may begin to spread over large areas; crop yields may become variable, and the range of crops would be limited without irrigation. If large areas of tropical rainforest are replaced by grassland, there is likely to be much less evapotranspiration and rainfall during each year in these areas than is currently experienced, which will lead to reduced cloud cover and surface air temperatures that are higher. Changes will occur in albedo and in energy and water balances (Charney *et al.* 1977). In all of the above studies, the effects of deforestation are concluded to be more devastating than was previously thought to be the case, extending to the regional and even global levels (Clark 1992: 47; Sellers *et al.* 1986).

Among the most far-reaching and immediate effects of deforestation is the modification of micro-climates (Meher-Homji 1991: 163). Whereas forests do not seem to influence cyclonic or orographic types of rainfall, they seem to influence convectional rainfall. It may not be the absolute decline in monsoon rainfall that affects agriculture, flora and fauna, and water supply, but rather the lack of rains at critical stages. However marginal may be the increase in rainfall due to forest cover, it makes a difference in sustaining agricultural crops and maintaining ecosystems. The determinant factor is not only the vegetation but also the relationship between the moisture content of soil, vegetation and the solar energy necessary for transforming water into atmospheric water vapor.

Carbon sequestration

The tropical forests of the world cover 11.5 per cent of the earth's land surface, yet they contain 46 per cent of the living terrestrial carbon (Brown and Lugo 1982). Measurements of carbon dioxide (CO_2) concentrations in air bubbles trapped in Greenland and Antarctic ice show that concentrations were stable near 280 ppm for at least 1,000 years before the more recent ongoing exponential increase in CO_2 began (Neftel *et al.* 1985; Watson *et al.* 1990). Further measurements of CO_2 concentrations have been carried out since 1957 at which concentration levels of CO_2 were 315 ppm; during that time, concentrations have increased to over 360 ppm at present, and the rate of increase has accelerated (Houghton 1991; Keeling *et al.* 1989; Tucker 1995: 215). About a third was thought to have come from deforestation (Houghton and Skole 1990).

Although the emissions are increasing over time, if humans choose, the process is reversible. According to Houghton (1991) absorption of atmospheric CO_2 by the oceans could remove about 85 per cent of the emitted CO_2 within a few hundred years; dissolution of marine carbonate sediments could remove another 10 per cent or so within a few thousand years; and silicate weathering could take care of the rest within about 100,000 years, which is a very short period of time from an evolutionary perspective.

The role of forests was given great prominence with the signing of the Kyoto Protocol of the United Nations Framework Convention on Climate Change in 1997. Globally, carbon sequestration from reduced deforestation, forest regeneration and increased development of plantations and agroforestry between 1995 and 2050 could amount to 12 to 15 per cent of fossil fuel carbon emissions (Houghton 1991: 112). However, the necessary forest management strategies depend in large part on the implementation of the Kyoto Protocol. Countries would be required to report deforestation, promote sustainable forest management practices, and develop, through afforestation and reforestation, "sinks" for purposes of sequestration.

Unfortunately, the debate continues to rage between poor and rich nations as to responsibilities under the Kyoto Protocol. In November 1998, negotiators, policy-makers and environmental activists convened in Buenos Aires for the Fourth Conference of the Parties to the United Nations Framework on Climate Change. The delegates from 160 countries met to work out the terms of the Protocol, in which governments set 2012 as the deadline for cutting back on greenhouse gases. Key issues of land use and forestry were covered. Controversy marked the event. Industrialized nations led by the United States insisted that developing nations must make a concerted, voluntary effort to reduce harmful carbon emissions. Given that the industrial nations have accumulated around 86 per cent of the world's emissions between 1870 and 1986, and that the United States is responsible for around a quarter of CO_2 emissions worldwide, the response of the developing nations should not be a surprise. For that matter, there is some validity for the case put by the United States and Japan. Necessary reduction of carbon output will require major structural shifts in the production of goods and services by industrial nations, with energy price increases, increases in structural unemployment, and declines in annual GDP. The fear also is that companies will simply go offshore to nations who are not, or have no desire to reduce economic output by adhering to standards, voluntarily or involuntarily.

Biodiversity

Biodiversity means "the genetic, taxonomic, and ecological variability among living organisms; this includes the variety and variability within species, between species, and of biotic components of ecosystems" (United Nations 1992: Chapter 15). Diversity can be measured at different levels (genes, species, higher taxonomic levels, communities and biotic processes, and ecosystems or ecosystem processes) and at different scales (temporal and spatial). The measurement can be calculated in number or relative frequency (Redford and Robinson 1995: 402).

Forests are dynamic, interactive structures that respond to natural disturbances such as drought, disease or a change in climate. Chaos is the norm and change is the consequence. Old growth forest is born of catastrophe. While most debates focus on the "charismatic species" that may, or may not, be vital in the functioning of the forest, hundreds of species of arthropods, from insects and spiders to centipedes and beetles, are overlooked. At the bottom of the heap are the fungi and microorganisms. The underrated forms of life are vital to the health of the forest (Dayton 1990: 23–24).

Genetic diversity refers to the variability within a particular species, population, variety, subspecies, or breed, as measured by the variation in genes (chemical units of hereditary information that can be passed from one generation to another). All genetic diversity ultimately arises at the molecular level, based on the properties of nucleic acids. There is no single way to measure genetic diversity, but it can be assessed by DNA and protein polymorphism as well as by detection of polymorphism in quantitative morphological traits (Bawa *et al.* 1991).

Species diversity refers to the variety of living species on earth. It can be measured at a local, regional, or global scale in a number of different ways, which differentially weight presence vs. frequency of different species at a given locality. The species is the unit best understood, and the one that biologists most commonly use to categories the variation of life. As a result most attention on biodiversity has been focused at the species level. There is, however, considerable disagreement on how to define a species and how to measure the diversity of species. Diversity of higher taxonomic levels (genera, families, etc.) refers to the variety of organisms within a given region at a taxonomic level higher than the species level (Redford and Robinson 1995: 402).

Communities and biotic process diversity refers to the variety at the level of a group of organisms belonging to a number of different species that co-occur in the same habitat or area and interact through trophic and spatial relationships. Biotic process diversity includes processes such as pollination, predaciousness, and mutualism (Hunter *et al.* 1988). Ecosystems-level diversity refers to the variety of communities of

organisms and their physical environment interacting as an ecological unit. It can be further divided into ecosystem types and ecosystem processes. Types are bounded communities interacting with the abiotic environment. Ecosystem processes can be classified on the basis of:
- functional attributes, that is, the capacity of the ecosystem to capture, store, and transfer energy, nutrients, and water; or
- structural attributes, relating to abundance and distribution of species of various sizes and shapes, such as species as structural types; or
- functional types, referring to the abundance and distribution of species with such functional attributes as the capacity to fix nitrogen and behave as predators or pollinators (Andersen *et al.* 1991).

At the ecosystem level conservation is in large part conservation of properties and processes, not of species or assemblages of species, because of the substitutability and redundancy of species within an ecosystem (Ehrlich and Mooney 1983; Walker 1992). Fundamentally, it is the resilience and capacity of the ecosystem to adapt to change, rather than sustainability, that is at issue.

Although extinction is a natural phenomenon, under normal circumstances, an average species lasts perhaps 10 million years from its appearance to extinction (Ehrlich and Wilson 1991). "Normal" conditions are punctuated by episodes of mass extinction, of which five are known in the past hundreds of millions of years. Presently, it appears that human activity is dramatically accelerating the process of another extinction phase (IUCN 2000). A number of observers have made the point that we are into the opening phase of a severe species-extinction spasm. A total of 11,046 species of plants and animals are threatened, facing a high risk of extinction in the near future, in almost all cases as a result of human activities. Indonesia is among the countries with the most threatened mammals and birds, while plant species are declining rapidly throughout Southeast Asia, particularly in Malaysia (IUCN 2000). Evolutionary response will eventually come up with replacement species in numbers and variety to match today's array, but the time required is measured in millions of years (Ehrlich and Ehrlich 1981; Jablonski 1991; Myers 1993; Raup 1991a; Soule 1991; Western and Pearl 1992). The loss of diversity is by far the least reversible component of global change (Randall 1986; Vitousek and Lubchenco 1995: 61).

Species are being lost in other biomes as well, notably coral reefs, wetlands, islands and montane environments. However, all of these other areas put together do not remotely match tropical forests in terms of numbers, but they push the estimated current extinction total beyond

30,000 species per year (Raup 1991b). The tropical rainforests of Asia seem to be most endangered in terms of likely impact of deforestation on species richness. The estimated loss is twice, though the area deforested is about half that of Latin America and the Caribbean (FAO 1991: 24). A number of independent analyses propose that we face the prospect of losing 20 per cent of all species within 30 years and 50 per cent or more by the end of the 21st century. All these estimates are explicitly conservative. Whitten *et.al.* (1987: 487), on the assumption that 10,000 ha of lowland forest are permanently cleared, starkly quantifies the kill rate. The following would be among the subsequent deaths: 30,000 squirrels; 5,000 monkeys; 15,000 hornbills; 600 gibbons; 20 tigers; and 10 elephants.

Economic arguments supporting biodiversity conservation in national parks, reserves and forests are widely publicized (Dixon and Sherman 1990; McNeely *et al.* 1990; Braetz 1992). Of concern, based on recent African research, is that the net revenues from alternative non-timber sources of revenue are unlikely to meet the opportunity costs of the land set aside in parks, reserves and forests for decades to come—if ever. In effect, therefore, a nation such as Kenya is subsidizing conservation activities, of benefit to the world, to the amount of 161 million USD annually. The chief values of these conservation activities are all indirect and external. Very few Kenyans visit parks, reserves and forests: further, many of the indirect values of conservation, such as wildlife experience, existence values, biodiversity values and carbon sequestration, are also external to Kenya (Norton-Griffiths and Southey 1995: 135–137).

The "precautionary principle" is especially relevant to the biodiversity problem (Cameron and Abouchar 1991; Costanza and Cornwell 1992; Perrings 1991). The principle asserts that there is a premium on a cautious and conservative approach to human interventions in environmental sectors that are (a) unusually short on scientific understanding, and (b) unusually susceptible to significant injury, especially irreversible injury. Given the biological, ecological, genetic, evolutionary, economic, aesthetic and ethical reasons for us to regret the loss of any species, the burden of scientific proof should remain the responsibility of environmental disrupters (Ehrlich and Wilson 1991; Morowitz 1991; Norton 1987).

Neoclassical Economic Paradigm

Conventional neoclassical economics is essentially concerned with the tropical rainforest as a renewable resource (Simmons 1993: 18–45). The maintenance of constant real consumption expenditure over time (maximum sustainable income) (Hicks 1946) requires the maintenance of the value of the asset base, which includes natural resources (Solow 1986). Whether or not renewal or destruction of a rainforest is a benefit or loss within this paradigm becomes a province of calculation, i.e., cost-benefit analysis (CBA). CBA weighs the costs of proceeding with a strategy as opposed to the benefits that could be derived from it. In order to carry out this procedure, monetary values are assigned to identifiable costs and benefits, both those which have commercial values found in the market, as well as an "extended utilitarian accounting" of those found within the human preference structure but not exchanged in the market (Beder 1993; McNeely 1988: 1–36).

The philosophical premises of CBA are utilitarian, anthropocentric and instrumentalist: "utilitarian in that things count to the extent that people want them; anthropocentric, in that humans are assigning the values; and instrumentalist, in that biota is regarded as an instrument for human satisfaction" (Randall 1988: 218). Two powerful epistemological assumptions lie at the heart of this microeconomic perspective. First, that decisions concerning the use of natural resources are typically made by private individuals or corporations seeking to maximize their own welfare through the operation of competitive markets; and second, that only this process will ordinarily yield allocations of resources approximating the social optimum. Given these assumptions, "effective systems of management can ensure that biological resources not only survive, but in fact increase while they are being used, thus providing the foundation for sustainable development" (McNeely 1988: 2). Should the biological resources not survive, sustainable development is still possible if the value derived from those resources is put to use providing utilitarian alternatives for the original resource. In this way neoclassical economists are able to handle the problem of intergenerational inequity by arguing that future generations will not know what they are missing should the original resource no longer be available, and in any case, will have a viable utilitarian alternative for the resource they no longer possess.

CBA is often able to provide useful information for the public policy process. Nevertheless, rainforest degradation is among those problems for which the impediments to high-quality CBA are greatest. The flaw in the CBA approach, as is true for much of social and behavioral science, is that when it comes to theories of complex phenomena, making phenomena more analyzable may be convenient, but does not necessarily make the analysis better at describing what is actually occurring (Gell-Mann 1995:

323). The interdependency of ecological communities and ecosystems indicates that any calculations of losses within an ecosystem in the present must take into account the loss of future benefits from other ecosystems and ecological communities that also succumb. Therefore, in order to calculate the value of a particular ecosystem, the values of all other ecosystems and ecological communities would somehow have to be determined. In theory, ecological information could be factored into these calculations, "but of all the areas of biology and ecology, few are less understood than the interspecific dependencies " (Norton 1988: 203).

Finally, philosophical individualists generally oppose CBA's lack of concern for non-consensual and uncompensated losses to some individuals and the greater gains it awards to others; and, the simple aggregation of individual valuations which implicitly accepts the society's existing distribution of endowments as well as implicitly denying any form of public interest apart from the summation of private valuations. The decision rule of CBA, that is, only the sum of the value gained (benefits) by beneficiaries exceed the sum of the value lost (costs) by others, does not provide security for the individual. CBA, as a universal criterion for collective decision-making, would subordinate individual rights for the overall good of the collective. Such a collective rule would justify government taking from the inefficient and giving to the efficient rather than the efficient purchasing from the inefficient (as in an exchange process). Within a market-based economy, this is an anomaly that makes neoclassical economists themselves discomfited.

Given this limited, but pertinent, review of the most well-used and powerful tool in the neoclassical economist methodology, the following section generates a synecological analysis of "production and distribution" within the context of the environment.

Synecological Relations of Production and Distribution

Fabricated within the analytical content of economic analysis is the unabashed proposition that "more is always better, irrespective of its distribution." The most basic example, learned by all undergraduate students in the discipline, is the general equilibrium postulate exemplified by an "Edgeworth box." The Edgeworth box didactically portrays a two-person, two-commodity world in which movement towards a contract curve provides for improvement in the efficiency of resource allocation, which is the goal of the exercise. As Daly has demonstrated

(1991: 36), while movement towards the contract curve enhances efficiency of allocation, moving along the curve leads to a change in distribution. The former is of concern to the economist, the latter considered to be outside the specifications of economics given the ethical connotations, but of relevance to policy-makers. Of equal concern however, in this context, is the fact that the scale of output is represented by the dimensions of the box, which are taken as given. Consequently, the issue of the optimal scale of the box itself escapes the limits of the analytical tool. The tacit answer to the implicit question seems to be that a bigger Edgeworth box is always better than a smaller one!

Marxism has responded in a number of ways to the challenge posed by ecology. There is much within its overall corpus, which is readily compatible with an ecological perspective, but it has to be reformulated and reconstructed. In fact, a Marxist anthropocentric approach has the main virtue of offering a reference point from which to evaluate ecological problems. Any "ecocentric" approach, on the other hand, is bound to be inconsistent, unless it adopts a mystical standpoint. It is inconsistent because it pretends to define ecological problems purely from the standpoint of nature. Many of the current custodians of the tropical rainforests have more pressing problems than the possibilities for curing cancer and AIDS, particularly when history suggests that the industrialized nations will be the major beneficiaries of their preservation (Vanclay 1993: 226). Unless one adopts a mystical or religious standpoint, there is always a human interest behind the attitude that nature should be left out there "for itself." The motives behind such a human interest may be of an aesthetic or a purely selfish character, or they may spring from humanity's general care about its environment (Grundmann 1991: 112–114).

One of the basic and still valid theses of Marx is that the form and the degree of growth of the forces of production assists in determining, but is also determined by, the social relations of production. Human social relations, of cooperation, competition or conflict, help to determine not only the human mode of production itself but also the integrated planetary ecological production system of sunlight/soil/water/energy/human labor (Omvedt 1992: 55). Nature is an integral part of any process of production. Because both social formations and nature are produced, they are malleable, transformable and transgressive (Smith 1984). An ecological movement that stands for the earth alone and ignores class and other social inequalities will succeed at best in displacing environmental problems, meanwhile reinforcing the dominant relations of power (Foster 1993: 12–13).

An ecocentric ethic, which inspires the sacrifice of humans to a higher morality, causes one to reflect on the American, General Westmoreland, who decided to wipe out a Vietnamese village in order to save it. Ethical positions themselves reflect class struggle and competition. Within a capitalist framework there is no reason for loggers to preserve old growth forests based on some morality beyond conscious beings. The metaphysical logic of "deep ecologists" espouses the changing of attitudes and values, without confronting or even understanding the material reality upon which these attitudes and values are based (Devall and Sessions 1985). Given their position that the preservation of old growth forests should not depend on human preference, one can only presume some enlightened mystical authority for their pronouncements (Godfrey-Smith 1980).

Within a market economy, a principle of distributive justice suggests that old growth should, in fact, not be preserved (Taylor 1986: 291–297). This was pointedly asserted recently by the Assistant Secretary of Agriculture for Resources and the Environment, John Crowell, when he said, "If you cut the old growth you're liquidating the existing inventory and getting the forests into a fully managed condition" (quoted in Foster 1993: 25). The economic justification for harvesting old growth forests is relatively simple. Old growth forests are decadent in the sense that increments to woody biomass, through growth, are offset by the death and decay of trees and tree branches. Also, old growth timber is highly valued because it is contained in large trees that produce high-quality, defect-free wood. Old growth forests thus contain large volumes of valuable wood, but are making no additions to total wood volume (Booth 1992: 45). From this standpoint, old growth forests should be harvested and converted to managed, even-aged forests that are harvested every 60–100 years. By leaving old growth forests standing, there will be no net production of new wood, the standing wood will go to waste, and we will forgo the opportunity to develop future timber flows from managed secondary growth stands (Pearse 1967). This type of instrumentalist ethic will always appear more sensible than one, which appeals to the morality of non-, or supra-, consciousness in the market.

Pursuit of the instrumentalist position makes sustainable rainforest management appear odd (Buschbacher 1990; Anderson 1990: 281), given that little else in developing countries works well, and when the agricultural systems and energy consumption of industrialized nations are clearly unsustainable. 27 countries with 97 per cent of the tropical forest owe one-half of the debt of less developed nations (Vanclay 1993: 230).

Even the most charitable interpretation of government policy would identify the incentive for government agents to cooperate with loggers to maximize the output of sawlog exports (Anderson 1989; Roselle and Katelman 1989). To suggest, as does Burgess (1993: 138), that deforestation is largely a function of "governmental failure" to use correct pricing and policy procedures, is at the least, a naive understanding of political economy, and at most, calculated disingenuousness.

Marx's conception of human emancipation included, in addition to the aim of freeing humans from coercive social relations and immediate physical want, the aim of fully developing and realizing human potential—affective, creative, aesthetic, spiritual, cognitive, etc., which also implies emancipation from the dictates of nature (Hayward 1992: 4). However, a materialist approach must recognize that nature is an integral part of the metabolism of social life. Social relations operate through the natural environment wherein both the social formation and nature are transformed and new socio-natural forms are produced. While nature provides the foundation, the dynamics of social relations produce nature's, and society's, history (Swyngedouw 1996: 68). Even, somewhat belatedly, international organizations have come to the realization that the inclusion of local peoples in conservation activities can increase the probability of success, as well as help reduce hostility and create the positive environment necessary to achieve the organizations' objectives (Kellert 1986; McNeely and Miller 1985). However, see Thompson (1999) for a more cynical view of this newly found penchant for participatory democracy.

Ecological degradation is an inherent part of the historically specific accumulation process that defines capitalist society and its class struggle (Foster 1993). This conception does not depend on viewing human potentialities as eternally pitted against nature, although that has been the view of orthodox Marxists in the past; but it nevertheless does see human potentialities as autonomous of natural causation, i.e., refusing to accept an unchanged essence conceptualized as "human nature." It is for this reason that the class context of reproducing social formations must be understood.

To pursue the intrinsic conflict involved in reproducing social formations through capitalist competitiveness, one can employ the micro and macro components of entropy (Rebane 1995: 89–90). The essence of the Second Law of Thermodynamics is that in the course of all processes in an isolated system the quality of energy and matter deteriorates—that after something has happened, their quality is lower than it was before (Atkins 1984). However, in part of the system, in subsystems, it is quite possible that in the process of growth of the larger system's entropy there appear ordered subsystems. The enhancement of orderedness, of negentropy, in a part of the system must take place at the expense of the deterioration of orderedness of the large system as a whole.

The ordered subsystems can live quite long, but they can be out of thermodynamical equilibrium only as long as the flows of high-quality energy in and of low-quantity energy out continue. This privileges the subsystem to survive and reproduce itself at the expense of other subsystems, which will have to vanish. Rebane (1995) makes the point cogently that it is the systems, which produce more entropy and, consequently, more pollution, or environmental degradation that are privileged to survive in the struggle for life. They will prevail because they act faster. The more active systems also take away high-quality energy and matter from the others, thereby suppressing their competitors. Therefore, while humans need the environment, evolutionary survival favors those systems, species, and communities and social formations that are the most active users (destroyers?) of nature.

Optimistic History of the Future?

> [When we say] if we know the present, then we can predict the future, it is not the consequence but the premise that is false. As a matter of principle we cannot know all determining elements of the present (W. Heisenberg cited in Wheeler and Zusek 1980: 62)

The FAO Global Forest Resources Assessment 2000 will be a most comprehensive and complete assessment to overcome the lack of information. It will consist of:
- A common set of definitions and standards for both tropical and temperate/boreal forests;
- A common set of parameters to be assessed, including not only the traditional timber information, but also information, where available, on non-wood forest products, naturalness, biological diversity, protected areas, carbon sequestration functions, forest condition and socio-economic functions;
- A remote-sensing-based global forest map keyed to forest classes used in the assessment, an ecoregion map and a database at 1km resolution;
- A global sample of satellite data to measure changes in forest cover over the past 20 years.

Demand for fuelwood is expected to continue to increase at a rate of about 1.1 per cent annually between now and 2010, while demand for industrial roundwood is expected to increase at a rate of about 1.7 per cent annually over the same period. However, given a more measured economic growth in the region, plus knock-on effects, overall consumption in the region across all wood product categories is estimated to be roughly 4 to 5 per cent lower than the baseline projection for the year 2010. Ironically, the Asian financial crisis has had the most unexpected and disruptive impact on forest product trade of any development in the 1997–98 period. A number of Asian currencies (Malaysia, the Philippines, Thailand and Indonesia) suffered significant depreciation, leading to economic recession. Given the dramatic changes in currency relativities and the extent to which Japan, the dominant importer, was affected by the recession, the effects of the crisis impacted on wood trade. There was an overall reduced demand for all forest products, affecting in particular, China, Japan, the Republic of Korea and Thailand; falling prices for forest products throughout the region; reduced earnings in the forestry sector; reduced harvests, workforce layoffs and the closure of mills (FAO 1999: 41).

Therefore, growth in the production of tropical forest products has slowed over the past three to five years. The export of most products has followed a similar trend; export volumes of tropical logs, sawnwood and wood-based panels have decreased. Most exports (by value) are from Asian countries, accounting for 80 per cent of tropical wood exports and more than 70 per cent of tropical wood imports by value. Japan alone accounts for about 40 per cent of tropical wood imports.

A by-product of the crisis was the International Monetary Fund's 1997 assistance package requirements to Indonesia. In return for assistance totaling about $40 billion USD, the IMF prescribed the following: an increase in the forest land tax, the abolition of APKINDO (the very powerful Plywood Association), establishing new resource rent taxes on timber resources, higher stumpage fees, and a transparent auction system to allocate concessions. In April 1998, the World Bank followed up with further reform requirements. These included linking forest royalties to world prices, reducing export taxes, incorporating local communities in the monitoring of forest resources, developing sustainable forestry land management targets, and a moratorium on the issuance of new logging licenses until the new measures are in place (Barber and Schweithelm 2000).

Overall, one could argue that the Asian financial crisis was a blessing in disguise for the tropical rainforests. It created a situation of lower demand and disrupted supplies. It may have also provided time for other protective and sustainable policies to be developed and put in place through international pressure and environmental necessity.

Given the environmental pressures, and catastrophes (floods, landslides, etc.) that have resulted from deforestation, some nations such as the Philippines have recently banned all logging in "old growth and virgin forests" and placed such forests under the National Integrated Protected Area System (Thompson 1997). The governments of China, Thailand and Cambodia have imposed bans of a similar nature. 54 member countries of the International Tropical Timber Organization (ITTO), accounting for more than 80 per cent of the world's tropical forests, and more than 95 per cent of the global timber trade, made a commitment to have their exports or tropical timber and tropical timber products come from sustainably managed sources by this year (FAO 1999: 14). A new fund has been established, predominantly subscribed to by the government of Japan, to assist member nations to implement sustainable forest management.

Certification and eco-labeling are proceeding,[3] although both remain controversial. Producer countries and trade groups highlight their trade-restrictive aspects; whereas consumer countries and environmental lobbies highlight the positive global externalities. Over the past two to three years the issues have been covered intensively and extensively at

Photo 8-1. Timber from a "certified" forest concession, Peninsular Malaysia. *Photo by Lye Tuck-Po.*

the Intergovernmental Forum on Forests in 1998 and 1999, the Intergovernmental Panel on Forests, UNCTAD, and the ITTO in Brazil in February 1999. Certification is primarily an issue for importing countries in Europe, but even Indonesia and Malaysia have put national certification systems into effect, if for no other reason than good public relations (FAO 1997b). While it appears that the process of certification will continue to expand, the evidence of its success remains to be shown. Most nations, of consequence, have developed or updated their National Forestry Plans during the past decade. The plans provide for not only enhanced silviculture and reduced inefficiencies, but also decentralization, recognition and respect for traditional and customary rights of the forest people, and more secure land tenure arrangements. Some governments (Papua New Guinea) are experimenting with semi-autonomous entities and private-public joint ventures. Others (Indonesia and Malaysia) are establishing partnerships with non-governmental organizations for purposes of research, extension and policy enforcement.

Finally technology and processing innovations are reducing demand pressures on the world's forests. It is expected that in most countries there will be a move away from the use of forest resources towards other land-based and non-land-based sources of supply. The greatest change by far will be the increased use of wood processing residues and recycled fibers in the product input mix. During the next decade, large areas of commercial short-rotation plantations (for both sawlogs and pulpwood) will begin coming on stream. Current trends in the use of recovered paper in the total fiber furnish will continue and improvements in harvesting practices will increase log recovery rates (FAO 1999: 56).

Conclusion

Political-economic analysis with attention to history makes clear that forests have been an essential context for human prosperity in wealthy countries. Just as clear is the point that this prosperity has come at a high price, both in terms of negative externalities and inadequate distributive justice. A straightforward economic vision of this process is inadequate because, with respect to ecological systems the market is often prone to failure. Human productive capacity requires sustaining the ecological balance, yet this is a theme ignored outside the market context of neoclassical economics.

Political ecology draws our attention to the fact that, not surprisingly, it is the relatively poor and weaker grassroots actors who bear a disproportionate share of the direct costs associated with local and regional environmental problems compared with their wealthier or more powerful counterparts. However, the former are often doubly disadvantaged in that they rarely receive a significant proportion of the benefits that usually attach to the economic activities that contribute to local or regional problems in the first place. In contrast, powerful actors derive their position in society in part from activities such as large-scale logging or cash-crop production that may be the main human contribution to such environmental problems. Therefore, while climate change, carbon sequestration and biodiversity draw the attention of the international media in terms of technical fixes, the work of political ecology is largely about seeking to explain the topography of a politicized environment.

The synecological relations of production and distribution remain well-informed through a Marxist analysis. We have to build on the Marxian argument that productive growth, as such, is not relevant to the abolition of poverty. What matters, always, is the way production is organized, who controls the work process, and the manner in which products are distributed. As presented above, nature is an integral part of the metabolism of social life and social relations operate through the natural environment. While nature provides the foundation for humanity, the dynamics of social relations produce both nature and social history.

The high expectations, once harbored, of development, a better material life, and increased political participation for the majority of people on this planet have scarcely been realized. This in itself should not surprise us, as over 40 years ago Claude Levi Strauss pointed out that historicist conceptions of progress are ultimately flawed: "…progress…is neither continuous nor inevitable; its course consists of leaps and bounds….These….are not always in the same direction" (1958: 21). In Asia, as in many other parts of the world, the problems are so staggering that conceptual understanding of the problems, not to speak of political and economic action to rectify them, increasingly seems beyond our capacities. Economists no longer advance theories of modernization or development with confidence. In fact, most have become inward looking to the macro and micro economic policies of the developed countries, leaving less developed countries and the majority of humankind to anthropologists and aid agencies (Gudeman 1992: 141). As Nobel Laureate A.K. Sen (1977: 336) put it succinctly: "…purely economic man is indeed close to being a social moron. Economic theory has been much preoccupied with this rational fool decked out in the glory of his one all-purpose preference ordering...we need a more elaborate structure."

Despite economists' claims that they are simply creating tools to examine an existing reality, the tools warrant the construction of one reality rather than another. Consequently, valuation models based on economics will pose considerable normative or ethical difficulties, which may provide good reasons to oppose the placement of dollar values on tropical rainforests. Cost/benefit analysis is largely limited by the implicit assumptions. First, one must accept that economists are able to control for all dimensions of quality. Second, one must assume that the non-market entity affects all people equally. Third, they must assume that there is no difference between the price a person is willing to pay to get something and the price she is willing to pay to avoid giving up something (Knetsch 1990: 227). Fourth, they must acknowledge that citizens do not differentiate between values expressed in private transactions and those expressed in public policy decisions. The incorporation of political ecology into our analysis of political economy confronts us with methodological fuzziness. For instance:

- Critical limits in the ecology of forests pose problems for any economic calculation of sustainability (Pearce *et. al.* 1989: 173–185; Barbier 1987; Lele 1991; Common and Perrings 1992; Harrison 1992; Opschoor and van der Straaten 1993);
- In economic decision models the notion of irreversibility, relevant to the preservation of biological diversity in rainforests, is contrary to the assumptions of infinite substitutability at the margin (Hodgson 1995);
- The "arrows of time" (Hawking 1988) are grounded in thermo-dynamics, whereas economists continue to presume technical replacement of resource depletion in both time and space, and;
- Evolution presents complexity and dissipative structures while economists cling to outdated presumptions of tendencies towards equilibrium (Anderson *et al.* 1988; Bak 1996).

National policy-makers are becoming more aware of the complex nature of policy reforms and the uncertainties created over wasteful forestry practices. The interrelationships between forests and other sectors of the economy are better understood. However, as usual, policy rhetoric means nothing without institutional capacity and reform. And institutional capacity and reform require the commitment of human and financial resources commensurate with the new responsibilities.

To productively contribute to the issue before us, we must learn to emphasize bounded rationality and the dynamics of evolution and learning. Instead of basing theory on assumptions that are mathematically convenient, we must construct models that are socially and psychologically explanatory. Instead of viewing the economy as a Newtonian machine, we must view it as something organic, adaptive, surprising, and alive. Instead of seeing the world as frozen in a black box of equilibrium and harmony, we must think about the world as an ever-changing system poised at the edge of chaos.

Acknowledgements

I wish to extend my appreciation for all of the arduous work by Abe-san, Lye Tuck-Po and Wil de Jong, who put in many hours to make separate contributions into a coherent collection. With reference to my own work I owe much to many, but I owe the most to Rosemarie, Dennise and Deanna.

Notes

1 Until the recent collapse of political order, the Solomon Islands was the world's sixth largest exporter of tropical hardwood logs, and forestry accounted for more than 50 per cent of its export revenues.
2 Dipterocarps are highly prized, light hardwoods reaching more than 45 m in height.
3 Eco-labeling refers to a guarantee that certain environmental standards have been met at all stages leading to the final product. It is broader in coverage than certification, which indicates that the product comes from forests managed according to a defined environmental standard.

Bibliography

Agarwal, A. and S. N. 1991. *Global Warming in an Unequal World*. New Delhi: Center for Science and Environment.
Anderson, A.B., ed. 1990. *Alternatives to Deforestation: Steps toward Sustainable Use of the Amazon Rain Forest*. New York: Columbia University Press.
Anderson, P. 1989. The Myth of Sustainable Logging: The Case for a Ban on Tropical Timber Imports. *The Ecologist* 19: 166–168.

Anderson, P., K.J. Arrow and D. Pines. 1988. *The Economy as an Evolving Complex System.* Redwood City, CA: Addison-Wesley.

Andersen, R., E. Fuentes, M. Gadgil, T. Lovejoy, H. Mooney, D. Ojima, and B. Woodmansee. 1991. Biodiversity from Communities to Ecosystems. In O.T. Solbrig, ed., *From Genes to Ecosystems: A Research Agenda for Biodiversity,* 73-82. Cambridge, Mass: IUBS.

Arrow, K., B. Bolin, R. Costanza, P. Dasgupta, C. Fllke, C.S. Holling, B-O. Jansson, S. Levin, K-G. Mäler, C. Perrings and D. Pimentel. 1995. Economic Growth Carrying Capacity, and the Environment. *Science* 268: 520-521.

Atkins, P.W. 1984. *The Second Law.* New York: Scientific American Books.

Bak, P. 1996. *How Nature Works.* New York: Copernicus.

Barber, C.V. and J. Schweithelm. 2000. *Trial by Fire: Forest fires and Forestry Policy in Indonesia's Era of Crisis and Reform.* Washington, D.C. : World Resources Institute.

Barbier, E., J. Burgess, J. Bishop, B. Aylward and C. Bann. 1993. *The Economic Linkages Between the International Trade in Tropical Timber and the Sustainable Management of Tropical Forests.* London: London Environmental Economics Center.

Barbier, E.B. 1987. The Concept of Sustainable Economic Development. *Environmental Conservation* 14: 101-110.

——— 1993. Economic Aspects of Tropical Deforestation in Southeast Asia. *Global Ecology and Biogeography Letters* 3: 215-234.

Batabyal, A.A. 1995. Development, Trade, and the Environment: Which Way Now?. *Ecological Economics* 13(2): 83-88.

Bawa, K., B. Schaal, O.T. Solbrig, S. Stearns, A. Templeton, and G. Vida. 1991. Biodiversity from the Gene to the Species. In O.T. Solbrig, ed., *From Genes to Ecosystems: A Research Agenda for Biodiversity,* 15-36. Cambridge, Mass: IUBS.

Beckerman, W. 1992. Economic Growth and the Environment: Whose growth? Whose environment? *World Development* 20: 481-496.

Beder, S. 1993. *The Nature of Sustainable Development.* Victoria, Australia: Scribe Publications.

Booth, D.E. 1992. The Economics and Ethics of Old-Growth Forests. *Environmental Ethics* 14(1)(Spring): 43-62.

Borhan, M., B. Johari and E.S. Quah. 1990. Studies on Logging Damage Due to Different Methods and Intensities of Harvesting in a Hill Dipterocarp Forest Of Peninsular Malaysia. *Malaysian Forestry* 50(1/2): 135-147.

Braetz, S.,. 1992. Conserving Biological Diversity: A Strategy for Protected Areas in the Asia-Pacific Region. *World Bank Technical Paper 193.* Washington, D.C.: IBRD.

Brookfield, H., L. Potter and Y. Byron. 1995. *In Place of the Forest: Environmental and Socio-economic Transformation in Borneo and the Eastern Malay Peninsula.* Tokyo; New York; Paris: United Nations University Press.

Brown, S. and A.E. Lugo. 1982. The Storage and Production of Organic Matter in Tropical Forests and Their Role in the Global Carbon Cycle. *Biotropica* 14: 161–187.

Bruijnzeel, L.A. 1992. Managing Tropical Forest Watersheds For Production: Where Contradictory Theory And Practice Co-Exist. In Miller, F., ed., *Wise Management of Tropical Forests*, 37–75. Oxford: Forestry.

Burgess, J.C. 1993. Timber Production, Timber Trade and Tropical Deforestation. *Ambio* 22(2/3)(Ma): 136–143.

Buschbacher, R.J. 1990. Natural Forest Management in the Humid Tropics: Ecological, Social, and Economic Considerations. *Ambio* 19(5): 253–258.

Cameron, T. and J. Abouchar. 1991. The Precautionary Principle: A Fundamental Principle of Law and Policy for The Protection of the Global Environment. *Boston College International and Comparative Law Review* 14: 1–27.

Cannon, C.H., D.R. Peart, M. Leighton and K. Kartawinata. 1994. The Structure of Lowland Rainforest after Selective Logging in West Kalimantan, Indonesia. *Forest Ecology and Management* 67: 49–68.

Charney, J.C., W.J. Quirk, S.W. Chow and J. Kornfield. 1977. A Comparative Study of the Effect of Albedo Change on Drought in Semi-Arid Regions. *Journal of Atmospheric Science* 34: 1366–1385.

Chichilnisky, G. 1994. North-south Trade and the Global Environment. *American Economic Review* 84: 851–874.

Clark, C. 1976. *Mathematical Bioeconomics: The Optimal Management of Renewable Resources.* New York: John Wiley.

——— 1992. Empirical Evidence for the Effect of Tropical Deforestation on Climatic Change. *Environmental Conservation* 19(1) (Spring): 39–47.

Common, M. and C. Perrings. 1992. Towards an ecological economics of sustainability. *Ecological Economics* 6: 6–34.

Costanza, R. and L. Cornwell. 1992. The 4-P: approach to dealing with scientific uncertainty. *Environment* 34: 12–17.

Daly, H. 1991. Elements of Environmental Macroeconomics. In Costanza, R., ed., *Ecological Economics: The Science and Management of Sustainability*, 32–46. New York: Columbia University Press.

Dargavel, J. and R. Tucker, eds. 1992. *Changing Pacific Forests: Historical Perspectives on the Forest Economy of the Pacific Basin.* Durham, North Carolina: Forest History Society.

Dayton, L. 1990. New Life for Old Forest. *The New Scientist* 13: 21–25.

Devall, B. and G. Sessions. 1985. *Deep Ecology: Living as if Nature Mattered.* Salt Lake City, Utah: G.M. Smith.

Dixon, J.A. and P.B. Sherman. 1990. *Economics of Protected Areas: Approaches and Applications.* Honolulu: East-West Center.

Ehrlich, P.R. and A.H. Ehrlich. 1981. *Extinction.* New York: Random House.

Ehrlich, P.R. and H.A. Mooney. 1983. Extinction, Substitution, and Ecosystem Services. *Bio Science* 33(4): 248–54.

Ehrlich, P.R. and E.O. Wilson. 1991. Biodiversity Studies: Science and Policy. *Science* 253: 758–62.

FAO (Food and Agricultural Organisation). 1989. *Household, Food Security, and Forestry: An Analysis of Socio-Economic Issues.* Rome: FAO.

———— 1991. Forest Resources Assessment in Tropical Countries, 1990. *FAO Forestry Paper* No. 112, Rome: FAO.

———— 1997a. *Regional Study—the South Pacific.* Working Paper APFSOS/WP/O1, Rome: FAO.

———— 1997b. *State of the World's Forests.* Rome: FAO.

———— 1998. Reduced Impact Timber Harvesting in the Tropical Natural Forest in Indonesia. Forest Harvesting Case-Study No. 11, Rome: FAO.

———— 1999. *State of the World's Forests.* FAO: Rome

Far Eastern Economic Review. Various issues.

Foster, J.B. 1993. The Limits of Environmentalism Without Class: Lessons from the Ancient Forest Struggle of the Pacific Northwest. *Capitalism, Nature, Socialism: A Journal of Socialist Ecology.* 4(1): 11–42.

Gell-Mann, Murray. 1995. *The Quark and the Jaguar: Adventures in the Simple and the Complex.* London: Abacus.

Godfrey-Smith, W. 1980. Environmental Philosophy. *Habitat* 8(3): 24–5.

Godoy, R. and R. Lubowski. 1992. Guidelines for the Economic Valuation of Nontimber Tropical-Forest Products. *Current Anthropology* 33(4): 423–432.

Goodland, R.J.A., E. Aisley, J.C. Post, and M.B. Dyson. 1990. Tropical Forest Management: The Urgency of Transition to Sustainability. *Environmental Conservation* 17(4) (Winter): 303–318.

Gowdy, J.M. and C.N. McDaniel. 1995. One World, One Experiment: Addressing the Biodiversity-Economics Conflict. *Ecological Economics* 15(3): 181–192.

Grundmann, R. 1991. The Ecological Challenge to Marxism. *New Left Review* No. 187: 103–120.

Gudeman, S. 1992. Remodeling the House of Economics: Culture and Innovation. *American Ethnologist* 19(1): 141–154.

Harrison, P. 1992. *The Third Revolution: Population, Environment and a Sustainable World*. Harmondsworth: Penguin.

Hawking, S.W. 1988. *A Brief History of Time: From the Big Bang to Black Holes*. London: Bantam.

Hayward, T. 1992. Ecology and Human Emancipation. *Radical Philosophy* 62 (Autumn): 3–13.

Henderson-Sellers, A. and V. Gornitz. 1984. Possible Climatic Impact of Land Cover Transformations, with Particular Emphasis on Tropical Deforestation. *Climatic Change* 6: 231–57.

Hicks, J.R. 1946. *Value and Capital*. Oxford: Oxford University Press.

Hodgson, G.M. 1995. *Economics and Biology*. Cheltenham: Elgar.

Houghton, R.A. 1990. The Global Effects of Tropical Deforestation. *Environment, Science & Technology* 24(4): 414–422.

———— 1991. Tropical Deforestation and Atmospheric Carbon Dioxide. *Climatic Change* 19(1/2): 99–118.

Houghton, R.A. and D.L. Skole. 1990. Carbon. In B.I.Turner, W.C. Clark, R.W. Kates, J.F. Richards, J.T. Mathews, and W.B. Meyer, eds., *The Earth As Transformed by Human Action*, 393–408. Cambridge: Cambridge University Press.

House of Commons Environment Committee. 1991. *Climatological and Environmental Effects of Rainforest Destruction*. London: HMSO.

Hunter, M.L., Jr., G.L. Jacobson Jr. and T. Webb. 1988. Palaeoecology and the Coarse-filter Approach to Maintaining Biological Diversity. *Conservation Biology* 2(4): 375–85.

IUCN (International Union for the Conservation of Nature). 2000. < http://www.iucn.org/> Accessed: 27 Oct. 2000.

Jablonski, K. 1991. Extinctions: a Palaeontological perspective. *Science* 253: 754–757.

Jacobs, M. 1988. *The Tropical Rain Forest: A First Encounter*. Edited by R. Kruk. New York: Springer-Verlag.

Kahn, J.R. and J.A. McDonald. 1990. *Third-World Debt and Tropical Deforestation*. Binghamton, New York: State University of New York Press.

——— 1995. Third-World Debt and Tropical Deforestation. *Ecological Economics*. 12(2): 107–123.

Keeling, C.D., R.B. Bacastow, A.F. Carter, S.C. Piper and T.P. Whorf. 1989. A Three Dimensional Model for Atmospheric CO_2 Transport Based on Observed Winds: Analysis of Observational Data. *Geophysical Monographs* 55: 165–236.

Kellert, S.R. 1986. Social and Perceptual Factors in the Preservation of Animal Species. In B. Norton, ed., *The Preservation of Species: The Value of Biological Diversity*, 50–73. Princeton: Princeton University Press.

Kerr, J.B. and C.T. McElroy. 1993. Evidence for Large Upward Trends of Ultraviolet-B Radiation Linked to Ozone Depletion. *Science* 262: 1032–1034.

Knetsch, J.L. 1990. Environmental Policy Implications of Disparities between Willingness to Pay and Compensation Demanded Measures of Values. *Journal of Environmental Economics and Management* 18: 227–237.

Lean, J. and D.A. Warrilow. 1989. Simulation of the Regional Climatic Impact of Amazon Deforestation. *Nature* 342: 411–413.

Leiss, W. 1974. *The Domination of Nature*. Boston: Beacon Press.

Lele, S. 1991. Sustainable Development: A Critical Review. *World Development* 19(6): 607–621.

Leslie, A. 1987. A second look at the economics of natural management systems in tropical mixed forests. *Unasylva* 155: 46–58.

Levi Strauss, C. 1958. *Race and History*. Paris: UNESCO.

Lewontin, R. and R. Levins. 1996. The Return of Old Diseases and the Appearance of New Ones. *Capitalism, Nature, Socialism: A Journal of Socialist Ecology* 7(2): 103–107.

Liu, S. 1996. Nature Conservation under Market Economic Conditions. *Ecological Economics* 1, (Spring): 41–44.

McNeely, J.A. 1988. *Economics and Biological Diversity: Developing and Using Economic Incentives to Conserve Biological Resources*. Switzerland: St. Mary's Press.

McNeely, J.A. and K.R. Miller, eds. 1985. *National Parks, Conservation, and Development: The Role of Protected Areas in Sustaining Society*. Washington, D.C.: Smithsonian Institution Press.

McNeely, J.A., K.R. Miller, W.V. Reid, R.A. Mittermeier and T.B. Werner. 1990. *Conserving the World's Biological Diversity.* Gland: IUCN.

Meher-Homji, V.M. 1991. Probable Impact of Deforestation on Hydro-logical Processes. *Climatic Change* 19(1/2): 163–173.

Monbiot, G. 1989. *Poisoned Arrows.* London: Abacus.

Morowitz, H.J. 1991. Balancing Species Preservation and Economic Considerations. *Science* 253: 752–754.

Munasinghe, M. and W. Shearer, eds. 1995. *Defining and Measuring Sustainability: The Biogeophysical Foundations.* Washington, D.C.: World Bank.

Myers, N. 1993. Biodiversity and the Precautionary Principle. *Ambio* 22(2/3): 74–79.

Nectoux, F. and Y. Kurida. 1989. *Timber from the South Seas: An Analysis of Japan's Tropical Timber Trade and its Environmental Impact.* Zurich: World Wildlife Fund.

Neftel A., E. Moor, H. Oeschger and B. Stauffer. 1985. Evidence from Polar Ice Cores for the Increase in Atmospheric CO_2 in the Past Two Centuries. *Nature* 315: 45–47.

Norton, B.G. 1987. *Why Preserve Natural Variety?.* Princeton, NJ: Princeton University Press.

——— 1988. Commodity, Amenity, and Morality: The Limits of Quantification in Valuing Biodiversity. In E.O. Wilson, ed., *Biodiversity*, 200–205. Washington, D.C.: National Academy Press.

Norton-Griffiths, M. and C. Southey. 1995. The Opportunity Costs of Biodiversity Conservation in Kenya. *Ecological Economics* 12(2): 125–139.

Omvedt, G. 1992. Fount of Plenty or Bureaucratic Boondoggle? India's Narmada Project. *Capitalism, Nature, Socialism: A Journal of Socialist Ecology* 3(4): 47–64.

Opschoor, H. and J. van der Straaten. 1993. Sustainable Development: An Institutional Approach. *Ecological Economics* 7: 203–222.

Ozanne, L. and P. Smith. 1993. Strategies and Perspectives of Influential Environmental Organizations Toward Tropical Deforestation. *Forest Products Journal* 43(4): 39–49.

Page, T. 1995. Harmony and Pathology. *Ecological Economics* 15(2): 141–144.

Panayotou, T., and P.S. Ashton. 1992. *Not by Timber Alone: Sustaining Tropical Forests Through Multiple Use Management.* Covelo, California: Island Press.

Pearce, D.W., A. Markandya and E.B. Barbier. 1989. *Blueprint for a Green Economy*. London: Earthscan.

Pearse, P.H. 1967. The Optimum Forest Rotation. *Forestry Chronicle* 43: 178–195.

Perlin, J. 1989. *A Forest Journey: The Role of Wood in the Development of Civilization*. New York: W.W. Norton.

Perrings, C. 1991. Reserved Rationality and the Precautionary Principle: Technological Change, Time and Uncertainty in Environmental Decision Making. In R. Costanza, ed., *Ecological Economics: The Science and Management of Sustainability*, 153–166. New York: Columbia University Press.

Randall, A. 1986. Human Preferences, Economics and the Preservation of Species. In Bryan G. Norton, ed., *The Preservation of Species: The Value of Biodiversity*, 79–109. New Jersey: Princeton University Press.

——— 1988. What Mainstream Economists Have to Say About the Value of Biodiversity. In E.O. Wilson, ed., *Biodiversity*, 217–223. Washington, D.C.: National Academy Press.

Raup, D.M. 1991a. *Extinction: Bad Genes or Bad Luck?* New York: W.W. Norton.

——— 1991b. A kill curve for Phanerozoic marine species. *Palaeobiology* 17: 37–48.

Rebane, K.K.

——— 1995. Energy, Entropy, Environment: Why Is Protection of the Environment Objectively Difficult? *Ecological Economics* 13(2): 89–92.

Redclift, M. 1993. Sustainable Development: Needs, Values, Rights. *Environmental Values* 2(1)(Spring): 3–20.

Redford, K.H. and J.G. Robinson. 1995. Sustainability of Wildlife and Natural Areas. In Mohen Munasinghe and Walter Shearer, eds., *Defining and Measuring Sustainability: The Biogeophysical Foundations*, 401–406. Washington, D.C.: World Bank.

Repetto, R. 1992. How to Account for Environmental Degradation. Paper prepared for the Conference on "Forestry and the Environment: Economic Perspectives," March, Alberta, Canada, University of Alberta.

Repetto, R., W. Magrath, M. Wells, C. Beer and F. Rossini. 1989. *Wasting Assets: Natural Resources in the National Income Accounts*. Washington, D.C.: World Resources Institute.

Roselle, M. and T. Katelman. 1989. *Tropical Hardwoods*. San Francisco: Rainforest Action Network.

Salati, E. and C.A. Nobre. 1991. Possible Climatic Impacts of Tropical Deforestation. *Climatic Change* 19(1/2): 177–196.

Sellers, P.J., Y. Mintz, Y.C. Sud and A. Dalsher. 1986. A Simple Biosphere Model (SiB) for Use within General Circulation Models. *Journal of Atmospheric Science* 43: 505–531.

Sen, A.K. 1977. Rational Fools: A Critique of the Behavioral Foundations of Economic Theory. *Philosophy and Public Affairs*. 16: 317–344.

Shukla, J., C. Nobre and P. Sellers. 1990. Amazon deforestation and climatic change. *Science* 247: 1322–1325.

Simmons, I.G. 1993. *Interpreting Nature: Cultural Constructions of the Environment*. New York: Routledge.

Sist, P., D. Dykstra and R. Fimbel. 1998. Reduced-impact Logging Guidelines for Lowland Dipterocarp Forests in Indonesia. *CIFOR Occasional Paper* 15. Bogor, Indonesia: CIFOR.

Smith, N. 1984. *Uneven Development: Nature, Capital and the Production of Space*. Oxford: Blackwell.

Smith, P.M., M.P. Haas and W.G. Luppold. 1995. An Analysis of Tropical Hardwood Product Importation and Consumption in the United States. *Forest Products Journal* 45(4): 31–37.

Solow, R.M. 1986. On the Intertemporal Allocation of Natural Resources. *Scandinavian Journal of Economics* 88: 141–149.

Soule, M.E. 1991. Conservation: Tactics for a Constant Crisis. *Science* 253: 744–750.

Swyngedouw, E. 1996. The City as a Hybrid: On Nature, Society and Cyborg Urbanization. *Capitalism, Nature, Socialism: A Journal of Socialist Ecology* 7(2): 65–80.

Taylor, P.W. 1986. *Respect for Nature: A Theory of Environmental Ethics*. Princeton: Princeton University Press.

The New Scientist. various issues.

Thompson, Herb. 1997. Philippine Forests: The Trees are Gone, Where's the Wood?. *Antepodium Electronic Journal* (September): 1–27. <http://www.vuw.ac.nz:80/atp/articles/thompson_9709.html>

────── 1999. Social Forestry: An Analysis of Indonesian Forestry Policy. *Journal of Contemporary Asia* 29(2): 187–201.

Thompson, H. and D. Kennedy. 1995. Cut now, pay later: Tropical rain forests of the Solomon Islands. *Journal of Mineral Policy, Business and Environment* 11(1): 21–29.

────── 1996. The Pulp and Paper Industry: Indonesia in an International Context. *Journal of Asian Business* 12(2): 41–55.

Tucker, M. 1995. Carbon dioxide emissions and global GDP. *Ecological Economics* 15(3): 215–223.

Turner, B.I., W.C. Clark, R.W. Kates, J.F. Richards, J.T. Mathews and W.B. Meyer, eds. 1990. *The Earth As Transformed by Human Action.* Cambridge: Cambridge University Press.

United Nations Conference on Environment and Development. 1992. *Conservation of Biological Diversity: Agenda 21.* Rio de Janeiro.

Vanclay, J.K. 1993. Saving the Tropical Forest: Needs and Prognosis. *Ambio* 22(4): 225–231.

Vitousek, P.M. and J. Lubchenco. 1995. Limits to Sustainable Use of Resources: From Local Effects to Global Change. In M. Munasinghe and W. Shearer, eds., *Defining and Measuring Sustainability: The Biogeophysical Foundations*, 57–64. Washington, D.C.: World Bank.

Walker, G. 1992. Diversity and Stability in Ecosystem Conservation. In Western, D. and Pearl, M.C. eds., *Conservation for the Twenty-first Century* , 121–130. Oxford: Oxford University Press.

Wallace, A.R. 1869. *The Malay Archipelago: The Land of the Orang-Utang and the Bird of Paradise.* London: Macmillan.

Wang, S. and Z. Xu. 1996. An Overview of Chinese Ecological Economics. *Ecological Economy* 1(Spring): 1–7.

Watson, R.T., H. Rodhe, H. Oeschger and U. Siegenthaler. 1990. Greenhouse Gases and Aerosols. In J.T. Houghton, G.J.Jenkins, and J.J.Ephraums, eds., *Climate Change: The IPCC Scientific Assessment*, 1–40. Cambridge: Cambridge University Press.

Western, D. and M. Pearl, eds. 1992. *Conserving Biology for the Next Century.* Oxford: Oxford University Press.

Wheeler, J. and W. Zusek, eds. 1980. *Quantum Theory and Measurement.* Princeton: Princeton University Press.

Whitmore, T.C. 1993. Changing Scientific Perceptions of the Eastern Tropical Rainforests: A Personal View. *Global Ecology and Biogeography Letters* 3: 115–121.

Whitten, A.J., S.J. Damanik, J. Anwar and N. Hisyam. 1987. *The Ecology of Sumatra.* Yogyakarta: Gadjah Mada University Press.

World Bank. 1991. *The Forest Sector.* Washington, D.C.: IBRD.

9
Discourse and Southeast Asian Deforestation: A Case Study of the International Tropical Timber Organization

Fred Gale

Explanations for the persistence of deforestation and forest degradation in tropical, temperate and boreal regions of the world occur at different levels of analysis. One approach treats agents—swidden farmers, fuelwood collectors, plantation owners, developers and loggers—as the root causes of forest destruction. This approach has the merit of focusing attention on what is taking place in a forest. Unless it is complemented by a broader contextual analysis, however, it can result in a form of "blaming the victim." Utting asks: "Are peasants at fault because they are clearing and burning large areas of forest or are they the victims of a particular socio-economic system which has made access to land and other resources in areas of greater agricultural potential increasingly difficult?" (1993: 14).

Standing back from specific agents and instances brings into view the role of state policy in deforestation and forest degradation. Attention shifts away from agents in the forest to politicians, bureaucrats and business executives in capital cities. Government failures in the form of perverse subsidies, outmoded legislation and inadequate resources are viewed as major explanatory variables. Market failures are also identified that include insecure property rights, monopolistic ownership patterns, and a lack of accurate pricing information through the timber chain. At this second level of analysis, the focus is on "better" state policies in the forest sector. What such better policies are, remains contested. In the current neo-liberal climate, however, powerful forces are promoting a smaller role for government and a larger role for the market.

Standing further back again, one can identify the important role scientific discourse plays in structuring debate and influencing policy development. Litfin notes: "As determinants of what can and cannot be thought, discourses delimit the range of policy options, thereby serving a precursors to policy outcomes" (1994: 37). Yet, discourses are not depoliticized. Litfin says that "discourses could not exist without individuals and groups promoting them, identifying with them, and even struggling with them. Discursive practices are inconceivable without discursive agents, coalitions, and knowledge brokers" (1994: 37–38). The dominant discourse for most of the 20th century in the field of forestry was sustained-yield forest management. It was displaced in the late 1980s by "sustainable forest management," an alterative discourse that developed at the International Tropical Timber Organization. At this third level of analysis, attention shifts outward from immediate agents and government policy to questions concerning the nature of discourse production and the adequacy of the discourse produced. We are now, however, a long way from the forest.

In this paper, the analytic emphasis is on the third, discursive level of analysis. The argument is that the International Tropical Timber Organization (ITTO) acted as a forum for a struggle over forest-management discourse. While that struggle led to a shift away from the "old forestry" paradigm of sustained-yield to a new discourse of "sustainable forest management" (SFM), the discursive shift was modest and the SFM discourse continued to legitimize many problematic features of the traditional approach. These included (a) viewing forests as "factories" for timber production; (b) requiring the calculation of an annual allowable cut (AAC); (c) vesting exclusive regulatory responsibility for forest management in the state; and (d) promoting forest exploitation through private enterprise and the market. The elaboration of the SFM discourse inhibited the development of a more profound discursive shift to ecosystem-based forest management, a shift that is still required if forests are to be sustainably managed.

The ITTO

The ITTO is the outcome of six years of negotiations sponsored by the United Nations Conference on Trade and Development (UNCTAD) under its 1976 Integrated Program for Commodities (IPC) (Cording 1985). At that time, and following the success of the Organization of Petroleum Exporting Countries (OPEC) in raising the price of oil on world markets, there was a perception that Third World states might be able to exercise commodity power over First World states by forming cartels. To head off this possibility and to achieve greater stabilization in primary product prices, UNCTAD under the leadership of Gamani Corea developed the Integrated Program for Commodities (UNCTAD 1977a). The objective of the IPC was to encourage the establishment of International Commodity Organizations (ICOs) in key resource areas such as tea, coffee, rubber, cocoa, tin and so forth. Tropical timber was a late addition to the list, but ultimately it became one of only a handful of commodities to successfully conclude negotiations under the IPC (for an overview of the history of ICOs, see Chimni 1987).

With reluctant endorsement by First World delegates, the IPC process commenced in earnest in 1977. Preparatory negotiating conferences were initiated for a number of commodities including tropical timber. The early negotiations were not auspicious, but effective background preparation in the UNCTAD Secretariat by Terrance Hpay and growing concern over poor management of tropical forests kept the momentum going (Hpay 1986). The early preparatory meetings focused on whether tropical timber was an appropriate commodity for an ICO. Key questions debated included: is tropical timber a "commodity"? Is the market for tropical timber segmented from the larger global timber market that includes temperate and boreal woods? By how much and why do tropical timber prices fluctuate? Should a buffer-stock in tropical timber be established to smooth out price fluctuations? How much are tropical timber producing and consuming countries prepared to pay to establish an international organization in the tropical timber sector (UNCTAD 1977b, 1977c, 1978)?

The preliminary discussions about the feasibility of an international commodity organization on tropical timber were completed in late 1980. From the perspective of many delegates, a case could be made to form an ICO in this resource sector. It would have to be tailored, however, to the unique features of the tropical timber market. Notably, agreement was reached (reluctant in the case of producer countries) on the non-feasibility of establishing a buffer-stock in tropical timber. In its place it was suggested that the ICO could be a vehicle for development assistance from North to South to improve Third World tropical forest management practices. In return for aid, producing countries would discuss their forest management policies, creating the possibility of more rapid progress on balancing the utilization of tropical forests with their conservation.

With the basis of a deal emerging, the Japanese government, whose industry depended heavily on tropical timber imports, took the initiative.

Japan produced an early draft of an international agreement on tropical forests, stimulating other countries to respond (UNCTAD 1981). The preparatory phase gave way to a negotiating conference in 1982 and by early 1983 delegates announced that they were ready to sign the International Tropical Timber Agreement. In addition to including in its objectives the goal of balancing tropical timber utilization with conservation, the ITTA-1983 also provided for the establishment of the International Tropical Timber Organization (ITTO) (ITTA 1984).

The ITTA-1983 was a complex agreement that aimed to utilize the tropical forests of the world to produce timber but to do so within limits to ensure their conservation. In addition to establishing the ITTO as a Secretariat to oversee and co-ordinate the Agreement's implementation, the ITTA-1983 also provided for three Permanent Committees (PCs) on Reforestation and Forest Management (PCF), Economic Information and Market Intelligence (PCM), and the Forest Industry (PCI). This structure—a small secretariat coordinating biannual meetings of delegates meeting simultaneously in plenary sessions and Permanent Committees—suggested that the organization would be largely member-driven rather than Secretariat driven. This structure contrasted with the largely Secretariat-drive Food and Agriculture Organization of the United Nations (FAO) (for more detail on the structure of the ITTO, see Gale 1998).

Voting arrangements at the ITTO reflected its history as a commodity agreement. Governments were divided into two categories: those who produced tropical timber and those who consumed it. Thus, Malaysia, Indonesia and the Philippines were considered tropical timber producers, while Japan, the Netherlands, and the United Kingdom fell within the consumer group. The North-South split implicit in this division did not work perfectly. Anomalous countries existed such as Thailand and Australia. The former is a net importer of tropical timber but a developing country; and the latter manages tropical rainforests in Queensland but is a developed country. These anomalies were resolved on the basis of North-South status, not on the basis of their producing and consuming status, an interesting bureaucratic sleight of hand.

Each of the two membership categories held a total of 1000 votes. The apportionment of these votes within each membership category differed. Four hundred of the producers' votes were first divided equally among the three continental groupings of Asia, Africa and South America, with the total regional votes being distributed equal to each country in the region. A formula allocated a further 300 votes based on the size of the tropical forest estate and a further 300 based on the value of their tropical timber exports. Notwithstanding this formula, African states agreed to allocate to each state an equal proportion of the votes assigned to the African group. This system of allocating votes within the producing-country group gave rise to three major states in terms of voting power: Brazil (largely based on forest size), Malaysia (largely based on export volume); and Indonesia (based on a combination of size and volume). Zaire, with a huge forest area, had the same number of votes as any other African country because of the internal African agreement to apportion votes equally.[1]

Consuming countries allocated ten votes to each member and then apportioned the remainder on the size of their imports of tropical timber. Import size was measured by volume of timber imports, not value. This overall formula benefited Japan, which held a huge number of votes at the outset.[2] The Japanese voting power was offset to some extent by the growing strength of the European Union (then the European Community). The rules of the EC/EU provided that member countries vote as a group when dealing with natural resources policy matters, but as independent states when dealing with matters of international development policy. Within the consuming-country group, then, meetings were first held between EC/EU members to agree a negotiating position on commodity matters. The EC/EU spokesperson then met with the head of the consuming country caucus to agree the consuming country position. It was only then that negotiations could occur between the producing and consuming countries to reach overall agreement. The institutional requirements of consensus within the EC/EU and consuming country caucuses ultimately weakened their negotiating position.

While much was initially made of the impact of the formal voting arrangements of the ITTO, in practice most of the ITTO's work was carried out through consensus bargaining within and between producing and consuming country caucuses. Voting power was important, however, for another reason. The number of votes a country held determined its contribution to the Organization's operating budget, which was the joint responsibility of both producing and consuming countries. Brazil, Malaysia, Indonesia and Japan contributed the lion's share of the Organization's annual budget. Since developing countries had to pay the assessed budget for the organization, and since financial resources for most of those countries were extremely tight, delegates were very parsimonious in making allocations to the ITTO's budget. Even today the Organization survives on an operating budget of only US$4.5 million, supporting a secretariat of some 30 staff (ITTO 1999: 15). The Organization would, indeed, be substantially smaller if it were not for generous additional donations from the Government of Japan and the City of Yokohama in the form of offices and staff on attractive terms.

One objective of the ITTO-1983, already stated, is to balance the utilization of tropical forests for timber production with their conservation. In attempting to operationalize this objective, the ITTO became the focus of attention of environmental organizations such as the International Institute for Environment and Development (IIED), the International Union for the Conservation of Nature (IUCN), the World Wide Fund for Nature (WWF) and Friends of the Earth (FoE). Indeed, the ITTO owed a debt to environmentalists whose intervention largely accounted for its coming into being. Central to this was the London-based IIED, which had carefully observed the closing sessions of the ITTA negotiations and, together with the IUCN, had been instrumental in inserting conservation language into the Agreement's objectives. In early 1985, with almost no ITTA ratifications submitted, it seemed that the ITTO would be still-born. The first ratification deadline had already passed and the fear in IIED and IUCN was that if the second deadline was not met there would be no possibility of the organization ever getting off the ground. In early March 1985, therefore, the IIED organized a conference on the ITTA and brought together key government and industry officials from Japan, the U.K., Malaysia, Indonesia, Brazil and the Netherlands to London to promote the organization and encourage ratification (for the proceedings of the conference, see IIED 1985). Subsequently, the IUCN, IIED and WWF used their extensive network of contacts to encourage governments to ratify the agreement. Japan, it appears, was active in promoting ratification with South American countries, especially Brazil.

The London Conference achieved its objective to the surprise of its principal architects. A sufficient number of treaty ratifications were received prior to the March 31 deadline to enable UNCTAD to convene the first meeting of the organization in June 1985. There, two outstanding matters required settling: the organization's location and the choice of its first Executive Director. These proved to be very contentious matters. It was accepted that if the location of the Secretariat was in a producing country, than the director should come from a consuming country and vice versa. Several cities were nominated for the Organization's location and several candidates were nominated for the Executive Directorship. No agreement could be reached at the first meeting, which adjourned and reconvened twice in November 1985 and July 1986.

Eventually, at the third convening of the first meeting of the ITTO deadlock was broken by the intervention of environmental organizations,

Photo 9-1. Pacifico Yokohama building where the ITTO secretariat is located.

which threatened to publicize the inter-governmental squabbling if a decision was not taken. Yokohama was chosen as the location of the new Secretariat and Freezailah bin Che Yeom, a Malaysian, was elected as its first Executive Director. The organization was empowered through a small budget to commence operations and its subsequent meetings were held in Yokohama in the autumn and in a producing country in the spring. The twice-yearly meetings placed cost and personnel demands on smaller countries that they had difficulty meeting and these countries were represented often by their local embassy. Consequently, the ITTO came to be dominated by government and industry interests from a small number of key timber producing and consuming countries. For producing countries, government and industry representatives from Brazil, Indonesia and Malaysia dominated discussions. For consuming countries, it was government and industry representatives from Japan, the Netherlands, the United Kingdom and the United States. It was within this structural context, then, that the organization commenced its work and set about achieving its objectives in the late 1980s.

The ITTO's Role in Southeast Asia

The ITTO has played both a general and a specific role in promoting "sustainable forest management" in Southeast Asia over the past 15 years. At a general level, the Organization contributed to the development of a discursive framework for a global forest regime. Much of this work in norm formation occurred in the late 1980s and early 1990s when the ITTO became, somewhat serendipitously, the only global forum in which debates about global forest policy could be actively engaged.[3] For example, one of the first activities of the Organization was to debate the need for and content of guidelines for "sustainable forest management" (SFM) of natural tropical forests. The ITTO was also the first global organization to consider the idea of forest eco-certification and eco-labeling, an area of immense importance in current debates on forest policy in the major timber exporting and importing states (Gale 2002).

But the ITTO's role in Southeast Asia's forests was also more direct and explicit. In 1989, in response to growing pressure from European and North American governments—pressure applied as a consequence of intense lobbying by European and American environmentalists allied with colleagues in producing countries, including Malaysia—the ITTO undertook a "Mission of Inquiry into the sustainability of forest management practices in Sarawak, Malaysia." This "Mission to Sarawak" was hugely controversial among producing country governments and the environmental community. Its prosecution and outcome resulted in the departure of environmental NGOs from the ITTO, which in turn damaged the legitimacy of the ITTO as a forum for global forest policy making.

To understand better the international context of Southeast Asian deforestation and forest degradation, this section focuses on the role played by the ITTO in three important initiatives. These are (a) the development and promotion of guidelines for the sustainable forest management of natural tropical forests; (b) the consideration of voluntary eco-certification and eco-labeling schemes; and (c) the conduct of its "Mission to Sarawak."

Guidelines for "Sustainable Forest Management"

At the start of the 1980s, when negotiations to establish an ITTA were being completed, there was no simple term for capturing what we now conceptualize as "sustainable development."[4] Consequently, the ITTO's objective that relates to this concept envisages the "sustainable utilization and conservation of tropical forests and their genetic resources" (ITTA 1984: Article 1 [h]). In pursuit of this objective, one of the first acts of the ITTO's newly appointed Executive Director was to let a contract to the International Institute for Environment and Development (IIED). The contract requested IIED to assess the degree to which a balance was being struck between the utilization and conservation of the world tropical forests. The consultancy was directed by IIED's Senior Forest Advisor, Duncan Poore (IIED 1988).

Poore's was one of the earliest studies to systematically consider the requirements of what has come to be known as sustainable forest management (SFM). His report conceptualized SFM as "the management of natural forests for the sustainable production of timber" (IIED 1988: 1) and embedded the concept firmly within the mainstream literature on forest management. His objective was to assess the degree to which forest management in tropical developing countries conformed to the requirements of sustained-yield forest management as it was practiced at the time in Europe, the United States, Canada and Australia. Indeed, the paradigm case of the management of natural forests for the sustainable production of timber was taken to be Queensland, Australia (IIED 1988: Regional Reports). Such an approach ignored growing criticisms of the "old forestry" literature and of the need to adopt a broader ecosystem approach to forest management by Maser (1988). It also ignored specific concerns about the adequacy of forest practices in Queensland (Keto,

Scott and Olsen 1990). Poore's study, despite a shallow ecological interpretation of SFM, revealed that "the extent of tropical forest which is being deliberately managed at an operational scale for the sustainable production of timber is, on a world scale, negligible" (IIED 1988: 2). In fact, although many countries had the stated intention of sustainable management, "progress in establishing stable sustainable systems is still so slow that it is having very little impact on the general decline in quantity and quality of the forest" (IIED 1988: 2). The study validated the findings of others. In Brazil, as Hecht and Cockburn documented, protection of the Amazonian forest was viewed as an impediment to enhanced national security and economic development via settlement and cattle ranching (Hecht and Cockburn 1989). In Indonesia, deforestation of East Kalimantan on the Island of Borneo was a product of a Suharto/World Bank sponsored transmigration program to resettle millions of Indonesians from heavily populated to underpopulated regions (SKEPHI and Kiddell-Monroe 1993).

Poore's report to the ITTO generated considerable interest and debate. Environmentalists attending the ITTO were delighted with a report that confirmed that tropically forested countries were not managing their forests sustainably. Some used the report to quietly lobby sympathetic government delegates at the ITTO to improve tropical forest management policies. Others employed the report as background material for campaigns designed to ban or boycott tropical timber imports. This created consternation among producing-country governments and business, who feared that overseas markets for tropical products would be curtailed.

A key recommendation of Poore's report was "the establishment of standards and manuals of 'best practice' for all the elements of management, harvesting and supporting research" (IIED 1988: 5). In response, the ITTO established a nine-member *ad hoc* Guidelines Working Group (GWG) composed of representatives from producing and consuming countries (three delegates each) and from intergovernmental, industry and environmental organizations (one delegate each). The terms of reference of the GWG was to prepare a draft set of forest management guidelines for discussion and ratification by the ITTO Council (Gale 1998: 146).

The GWG met face-to-face only once in London in February 1990. Most of its members were trained foresters, including the environmental representative from the World Wide Fund for Nature-International. Because GWG members had a similar formation, being trained by forestry schools in the practice of conventional industrial forestry, they conceptualized "good" forest management practices within a broadly similar scientific discourse. Members of the GWG were ably assisted by a consultant forester, Simon Rietbergen of the IIED, who relied heavily on Poore's report (Rietbergen 1990). In fact, the draft guidelines could be considered a systemization of the approach set out in Poore's 1988 IIED report (ITTO 1990a). The cost of the GWG was met by the U.K. government, which funded this ITTO pre-project activity.

Unsurprisingly, a working group composed almost exclusively of foresters developed a draft set of guidelines for "best practice" from within a conventional forestry paradigm based on the concept of "sustained yield." The emerging concept of SFM, developed by Poore and systematized into guidelines by the GWG, required governments to take a two-step approach. In step one, governments demarcated the total forest estate into two basic zones. The first, smaller forest zone was to be preserved in parks, nature reserves and protected areas for ecological and other forest values. The second, substantially larger zone was to be set aside as production forests and logged. This first step is set out in sections 2 and 3 of the ITTO Guidelines, notably section 2.3, Principle 7, which states: *"The different categories of land to be kept under permanent forest are...land to be protected; land for nature conservation; land for production of timber and other forest products; land intended to fulfill combinations of these objectives"* (ITTO 1992: 3).

In the second step, governments were obliged to "manage" their production forests according to sound silvicultural practices based on the concept of sustained yield. Necessary practices included demarcating forest boundaries, protecting forests against fires, disease and pests, using appropriate silvicultural systems (logging, replanting, spacing, thinning) and constructing good quality logging roads. The majority of the ITTO Guidelines deal with these practices, set out in detail in section 3 (ITTO 1992: 3–9).

With near consensus in the GWG on what constituted best practice in forestry, the only major bone of contention at the meeting was the social issue of the impact of parks and protected areas and logging on indigenous peoples' customary rights. Rietbergen's draft guidelines included language to protect indigenous peoples' rights but his clause was viewed as too demanding by other delegates, notably those from industry and producing countries.[5] The language concerning indigenous rights was weakened during the GWG negotiations and later further diluted at the 1990 Bali meeting of the ITTO Council, when the Report was formally considered.[6] Such an emasculation of the language on indigenous peoples' rights occurred because the GWG lacked a strong advocate to defend their interests. The homogenous composition of the GWG enabled the technical discourse of conventional forestry professionals on tree growing to dominate, a discourse which discounted the value of ecosystem theory, public consultation and local knowledge.

Thus it was that the ITTO was able to reach agreement extraordinarily quickly on a set of "best practice" guidelines that constituted "sustainable forest management" of the world's tropical forests. It took the GWG only two days to reach agreement on 99 per cent of the guidelines, the only subsequent changes being a weakening of the language on indigenous peoples rights. The GWG presented its report to the Eighth Meeting of the ITTO Council in May 1990. There, the draft guidelines were endorsed as written, with only one significant change. At the insistence of producing-country delegations, a phrase was inserted in the Introduction section to the Guidelines that read:

> These guidelines contain a set of principles, which constitutes the international reference standard established by the ITTO for the development of more specific guidelines, at the national level for sustainable management of natural tropical forests for timber production. The development, application and enforcement of national guidelines, based on this standard are matters for national decision by individual timber producing countries (ITTO 1992: 1).

The intent behind the insertion of this clause was to make it clear that consuming countries could not use producing-country failure to implement the ITTO guidelines as a justification for trade-restrictive measures. Producing-country governments were worried that some consuming-country governments might be able to justify the imposition of trade restrictions on tropical timber products on the basis that (a) there was an agreed international standard for SFM, which (b) was not being followed by producing country "x," which (c) justified trade sanctions. With the insertion of a "guidelines for guidelines" clause in the Introduction, producing-country governments could be held accountable only to their national guidelines not to ITTO's guidelines. In most cases, it should be noted, such national guidelines were not developed.

The Guidelines legitimized the status quo with respect to indigenous peoples, permitting the ongoing process of internal colonization of local resources in the interests of state development. Second, the ITTO Guidelines advocated an industrial approach to forest management, endorsing a shallow ecological and resourcist perspective. Even worse, perhaps, the Guidelines gave credence to the view that many producing countries—most notably Malaysia—were already practicing many elements of SFM. The message was that achieving SFM would require relatively minor adjustments from such countries. Finally, with the adoption of the guidelines-for-guidelines approach, even these weak and resourcist "best practices" to achieve SFM were rendered discretionary.

In summary, the ITTO acted as an inter-governmental forum to negotiate the discursive content of a new approach to forestry, one that legitimized an industrial conception of forest management. The discourse of SFM remains the hegemonic discourse at inter-governmental meetings on forestry matters today, notwithstanding significant developments in the science and practice of forestry in the past 20 years. According to a recent consultant's report, some foresters view the ITTO guidelines as outdated and passé; the majority, however, continue to view them as imperfect, but adequate.[7] A part of the explanation for continued deforestation and forest degradation in Southeast Asia lies in the concept of "sustainable forest management," the negotiated content of which perpetuates an industrial approach to forest use.

Forest Eco-Certification and Labeling

A second, less-than-progressive role played by the ITTO in the 1980s and 1990s concerns the use of voluntary market mechanisms to reassure consumers that the wood from which timber products are composed derives from forests that are sustainably managed. Ecocertification and ecolabeling schemes (ECLs) are important and interconnected means of achieving such customer reassurance. Yet, at the ITTO participating government and industry representatives sought initially to block debate on ECLs. Later, when such schemes began to be implemented anyway, the ITTO tacitly approved the adoption of weak national ECLs, which were viewed as a means of counteracting a more challenging scheme developed by environmentalists and institutionalized in the Forest Stewardship Council (FSC) (for details on FSC, see Gale 2000).

The history of ECLs at the ITTO bears out the above contentions. The story commences in 1988 when Koy Thompson, a forest campaigner with the Friends of the Earth-United Kingdom initiated discussions with Tim Synott of the Oxford Forestry Institute (OFI). Both were interested in placing a mark on tropical forest products that would signal that the wood came from sustainable sources. Ecolabels were becoming popular in the 1980s and Thompson and Synott wondered if they could be extended to the tropical timber industry.

Following exploratory discussions with officials at Britain's Overseas Development Administration (ODA) and with their encouragement, Thompson and Synott drew up a project proposal for a feasibility study on ecolabeling. The draft proposal was forwarded to ODA officials, who recommended minor, non-substantive changes to ensure the proposal conformed to the format of the funding organization, the ITTO. Subsequently, it was forwarded by the British government to the ITTO in 1989 for consideration as a pre-project proposal (ITTO 1989a). However, the proposal was greeted with a storm of criticism from producing-country governments and the producing and consuming arms of the tropical timber industry (for a detailed account, see Gale 1998: 158–177).

The objections raised to the proposal at the time were numerous. Producing-country governments backed by industry officials argued that eco-certification and labeling of tropical timber products would (a) act as a barrier to the trade in tropical timber products; (b) discriminate between tropical timber and timber from temperate and boreal forests; (c) be implemented unfairly because different countries would develop different standards; (d) subject the industry to unnecessary additional costs; and (e) be easily circumvented. In short, it was a waste of time.

Neither Synott nor Thompson, nor the British delegation that put forward the proposal were prepared for the stream of invective that the pre-project generated at the ITTO. The British delegation, led by a representative from the Department of Trade and Industry, quickly backed away from the proposal. The original draft was quickly and substantively reformulated at the meeting, and the offending eco-certification and labeling language was replaced with a feasibility study to explore the development of "incentives" to promote sustainable forest management (ITTO 1989b). While the language of incentives did not exclude eco-certification and labeling, it was evidently a broader and more all-encompassing notion. Subsequent careful handling by the ITTO Secretariat and British government ensured that the resulting consultancy, although granted to the Oxford Forestry Institute, was carried out by a team largely antithetical to the idea of ecocertification and ecolabeling. By these means, the Executive Director of the ITTO, producing country governments and the tropical timber industry aimed to kill off an unpalatable and potentially dangerous notion.

But the idea refused to die. In 1991, the OFI produced its report on "incentives" for sustainable forest management and endorsed a previously canvassed notion that a levy be placed on tropical timber imports by consuming countries with the funds generated used to support SFM in producing countries (OFI 1991). This proposal appealed to almost no one. Consuming-country governments found fault with the proposal because of the high administrative cost to collect and disburse the levy in an era of government downsizing, tight fiscal restraint, and perceptions of overtaxation. Producing-country governments feared that the levy would raise the price of tropical timber in importing countries, making them less competitive than substitute temperate and boreal timbers.

Between the letting of the OFI contract and its consideration by ITTO, an internal shift in the composition of the British delegation occurred. The Ministry of Trade and Industry handed over the lead role to the Overseas Development Administration, which meant that by 1991 ODA was once more in a position to back the idea of ecocertification and ecolabeling. Embarrassed by the OFI study and under pressure from its vocal environmental movement, the British Government decided to fund another study to subject the ECL notion to greater scrutiny.

The follow-up study was carried out by the London Environmental Economic Centre (LEEC), which presented a detailed theoretical and policy analysis of the contribution of the tropical timber trade to deforestation and the potential of incentives, including ECLs to ameliorate the damage done (LEEC 1993). The LEEC Report reflected a static conception of deforestation and ecocertification. Notably, the Report concluded that the trade in tropical timber contributed in a minor way to tropical deforestation and forest degradation. This conclusion was based on an analysis of the volume of wood entering international trade as a proportion of total production—a figure estimated in the Report to be approximately six per cent (LEEC 1993: iii). The LEEC Report concluded also that, to the extent that ECLs could be useful incentives to promote sustainable forest management, *country-level* schemes should be preferred over stand-level schemes (LEEC 1993: v).

Environmental organizations criticized the LEEC Report on its basic findings, especially on its preference for national ECLs. By this time, in early 1993, they were in a strong position to make such criticisms. The hostile reaction of producing-country governments and the tropical timber industry to their 1989 ITTO pre-project proposal had convinced environmentalists of the significance of the initiative. Delegates from WWF-International in particular were intrigued at the possibilities presented by ECLs and channeled money to explore the idea further. Several meetings were held among interested parties in the early 1990s to consider how to promote the idea. These meetings concluded that only stand-level certification based on international standards would work.

By early 1993, then, environmental organizations were in the final stages of creating a new international organization, the Forest Stewardship Council (FSC), when LEEC published its report (for background details on the formation of the FSC, see Dudley *et al.* 1995: 145–154).

Environmentalists objected strongly to the LEEC Report, therefore, because it played down the contribution of the timber industry to the destruction of the earth's tropical forests and because it backed government-sponsored, country-level not environmentally-sponsored, stand-level ECLs. Environmentalists pointed out that the authors of the LEEC report overlooked the dynamic contribution of the timber industry to tropical deforestation and forest degradation. Industry was responsible for far more damage than the size of its exports suggested because its activities opened up forests to exploitation by migrants, hunters, swidden agriculturalists and illegal loggers. Further, they argued, country-level certification schemes were unworkable. Such schemes gave national governments an incentive to develop weak standards for SFM, minimizing the degree of adjustment required by their forest products' industry. International pressure to harmonies the different schemes would result in lowest-common-denominator standards that would fail to halt tropical deforestation and forest degradation. Country-level ECLs would be open to abuse also since they would certify both scrupulous and unscrupulous operators. Given the high level of corruption in the industry—that included illegal cutting and evasion of royalty payments through misrepresentation of species and volumes cut—environmentalists argued that only stand-level certification would adequately prevent against fraud.

Environmental criticisms of the LEEC Report fell on deaf ears at the ITTO, but the efforts to establish the FSC were watched hawkishly by the industry. In 1993, as it became clear that the FSC would be set up, governments and industry associations one after another began to establish their own eco-certification and labeling processes to neutralize the FSC threat. In the United States, the American Forestry and Paper Association (AFPA) established its own voluntary Sustainable Forest Initiative (SFI) to certify its larger members. In Canada, the Canadian Pulp and Paper Association (CPPA) gave a million dollars to the Canadian Standards Association (CSA) to develop a made-in-Canada scheme. In Indonesia, industry, government and ECSOs collaborated to develop an Indonesian eco-label under a new organization, Lembaga Ekolabel Indonesia (LEI). And in the fall of 1993, the FSC was formally founded at a conference of delegates in Toronto, Canada (Dudley *et al.* 1995; Gale 2002).

The rapidity with which businesses and governments moved at the national level to blunt the FSC threat confirmed the power of the ECL idea. National action proved difficult to coordinate at the international level and governments participating at the ITTO were not able to respond collectively to new developments. In 1994, the ITTO considered a detailed report by consultants on the state of ECLs (Ghazali and Simula 1994). Subsequent updates were prepared and presented to the ITTO at two-year intervals (see Ghazali and Simula 1996, 1998; Nsenkyiere and Simula 2000). While these reports served to inform producing and consuming governments of ECL developments, the intergovernmental forum of the ITTO was unable to determine a common, never mind progressive, approach on the issue.

Why did it provide impossible to take action on ECLs at the ITTO? The basic explanation lies in the substantial disagreement ECLs provoked within and between ITTO caucuses. Some producing countries were in favor of eco certification (such as Indonesia, for example), but the majority were against (notably Malaysia). Some consuming countries were strongly in favor (notably the United Kingdom), many were lukewarm (such as France and Italy) and most favored schemes that would apply only to tropical timber imports. Some environmental organizations (such as WWF) strongly supported ECLs provided they operated at the stand-level. They were not in favor of country-level schemes. Other environmental organizations were more skeptical about the impact of ECLs and feared they might become techno-managerial solutions to the deforestation crisis. The confusion is well captured by a WWF Report on the ITTO debate on the LEEC document, which notes:

> The final report of the Market discussions [at which the LEEC Report was discussed]...turned into a procedural nightmare and finally a minimalist approach was taken. The reason for the problem was that Consumers were generally favorable to labeling but wanted it to apply to tropical timber only, whereas the Producers were more reticent, and would only support labeling if it applied to all timber (Elliott 1993: 3).

The ITTO Mission to Sarawak

Environmentalists supported the creation of the ITTO in 1985 out of growing concern for the state of the world's tropical forests based on evidence provided by the FAO, the World Resources Institute, the International Institute for Environment and Development and others. A parallel and partly over-lapping concern also emerged for the fate of the world's indigenous peoples, many of whom were forest dwellers. In the 1980s, however, many of the larger environmental organizations were not yet persuaded of the vital link between indigenous peoples' survival and rainforest protection. Smaller organizations, notably the Rainforest Action Network, Friends of the Earth, and Survival International, did work with indigenous peoples groups, however. One of their explicit roles was to make the link between indigenous peoples rights and rainforest protection more apparent to the larger, more conservative environmental groups.

One locale where the issue of indigenous peoples' rights and rainforest destruction became intertwined was in the Malaysian state of Sarawak on the island of Borneo. In 1987, members of the Penan, a society which traditionally were hunter-gatherers, blockaded roads in the Baram district to protest against logging in the region (WRM/SAM 1989; Brosius 1999). The Sarawak government responded by arresting several Penan and by placing the leader of Sahabat Alam Malaysia-Sarawak, Harrison Ngau, under house arrest. It also introduced legislation outlawing logging road blockades (WRM/SAM 1989; Sesser 1991). Sarawak authorities, together with the Malaysian federal government, launched a counterattack accusing "Northern environmentalists of 'eco-imperialism' and focused upon what was portrayed as the utter hypocrisy of Northern environmentalists for criticizing Malaysia's environmental and indigenous rights record" (Brosius 1999: 42).

As these events unfolded in Malaysia, the ITTO was commencing active operations. Environmental groups in Europe, already concerned about logging practices in Sarawak, intensified their campaigns for a boycott of tropical timber from Malaysia. Such boycotts were seen as effective levers to pressure the Sarawak government to rein in its logging industry. Environmental groups in Germany, the Netherlands and the U.K. were particularly effective in raising public awareness of the issues in Europe. In Malaysia, Sahabat Alam Malaysia received the alternative peace prize, the Right Livelihood Award, in 1988 in recognition of its efforts to assist the Penan (WRM/SAM 1989).

The importing and exporting arms of the tropical timber industry became increasingly alarmed at the negative publicity generated by environmental movements over the Penan issue. The Malaysian and Sarawak Governments were also deeply concerned. In 1989, the Malaysian government undertook a Mission to Europe to counteract the negative image of Sarawak forestry and reassure key importers that the Penan were being successfully and willingly resettled in villages to promote their economic development and integration into Malaysian society. The actions of a few renegade Penan, it was argued, should not derail policies that benefited a huge number of others (Government of Malaysia 1993).

While a Malaysian Mission traveled one way, a U.S. Mission to Sarawak traveled the other direction. Randy Hayes of the Rainforest Action Network organized a Congressional Staff Study Mission to Malaysia (CSSMM) in March/April 1989 (CSSMM 1989). Its subsequent report was a scathing attack on Sarawak's treatment of indigenous peoples and recommended *inter alia* that the United States "Review consumption patterns in the United States, look for ways of decreasing U.S. consumption of tropical hardwoods, and seek alternatives that could be substituted. Further, the U.S. should encourage consumer nations to reduce their consumption" (CSSMM 1989: 23). Other countries in Europe were agitating to carry out their own Missions to Sarawak and pressure was growing from many quarters, not least in Peninsular Malaysia itself, for a solution to the Sarawak "crisis." Producing country governments and industry feared that the growing campaign to boycott imports from Sarawak could be extended to tropical timbers from other states.

It was against this background that the ITTO's Malaysian Executive Director, Freezailah, traveled to Malaysia to solicit support for an ITTO inquiry into forest practices in Sarawak. While the details of Freezailah's visits and discussions remain confidential, it is evident that the initiative was his and that he lobbied the Federal and Sarawak state governments and industry representatives to permit the ITTO to undertake an inquiry. Freezailah's efforts bore fruit and at the May 1989 ITTO meeting, the Chief Minister of Sarawak announced that "the Government of Malaysia was ready to welcome a mission from ITTO to visit Sarawak to assist in promoting sustainable forest management" (ITTO 1989c: 2). His request led to a heated debate between Malaysian and South American delegates, with the latter objecting strenuously to the proposal for fear it would set a precedent for further, unsolicited visits to other producing countries.[8]

After detailed negotiations over the mission's terms of reference, consuming and producing country governments agreed that the Mission should proceed. The terms of reference specified that the Mission's purpose was "to assess the sustainable utilization and conservation of tropical forests and their genetic resources as well as the maintenance of the ecological balance in Sarawak, Malaysia." The administrative arrangements were handed over to the Executive Director who was empowered to "take all necessary measures for the implementation of this resolution and to prepare the necessary documentation for this purpose" (ITTO 1989c: 108–109). The Mission's terms of reference omitted mention of indigenous peoples' rights in Sarawak and did not include legal expertise to assess the degree to which existing rights were being enforced. This omission was notable given that it was the link between indigenous peoples' rights and logging that had created public pressure in Europe and the United States for the mission in the first place. It did not augur well for the Mission's "objectivity."

Following past precedent, it was agreed that a ten-member mission would be established to be led by an independent expert and to include three consuming and three producing country representatives, and one member each from ITTO's inter-governmental, industry and environmental organizations. By the end of the Council meeting the composition of the Mission had been determined. It was to be led by the Earl of Cranbrook and Duncan Poore of the IIED was to represent environmental interests. Although Cranbrook was not as "independent" as some would have liked, he was a naturalist with extensive experience in Malaysia and who spoke the language. Moreover, he had recently presided over a British House of Lords inquiry into Europe's policy on tropical forests and the timber trade and was thought by many to be a good compromise candidate.

The ITTO Mission to Sarawak took place in late 1989 and early 1990 and the delegation visited Sarawak three times. Cranbrook set up operational headquarters at the offices of the Royal Geological Survey in London and encouraged interested parties to contact him directly there. He thus preserved the integrity of the process and ensured that groups protesting against Sarawak's forest practices did not have to make their views known via the ITTO, the Sarawak or Malaysian authorities. Freezailah accompanied the Mission on each of its three visits to Malaysia, but neither he nor the Sarawak authorities directly constrained the Mission's agenda.

Since the Mission's mandate was to assess the degree to which sustainable forest management was being implemented in Sarawak and to make recommendations for its further strengthening, arguably the key element in its report was the estimate of Sarawak's Annual Allowable Cut (AAC). At the time of the Mission's visits to Sarawak, it was reported that the state's AAC was approximately 13 million m^3. By contrast, a detailed analysis by Castilleja of the U.S. National Wildlife Federation put the sustainable AAC at between 4 and 5 million m^3 (Castilleja 1990).

Cranbrook failed to generate a consensus among Mission members about the size of the AAC (Gale 1998). Some members argued that the existing AAC of 13 million m^3 was sustainable and could, perhaps, be expanded. Others, including Poore, argued that a significant reduction was necessary. A compromise arrangement was agreed whereby, following the mission members' final visit to Sarawak, they individually made their best estimate of the sustainable AAC and faxed their conclusions to Cranbrook in London. Cranbrook then selected a compromise figure based on these submissions and that reflected the diversity of submissions. His "guesstimate" was that the sustainable AAC was about nine million m,[3] which represented a substantial decrease from the current AAC. It was, however, considerably greater than the estimates of environmentalists, including Poore. Moreover, Cranbrook was careful to qualify the conditions under which even this AAC would be sustainable: these conditions included the use of special silvicultural techniques such as liberation thinning (ITTO 1990b).

The Mission's Report also, somewhat surprisingly, contained a wealth of information on Sarawak's system of customary land rights. At several points in the text, quasi-recommendations were made on how to improve indigenous peoples' access to the legal system. Despite the Mission's view that it could report and recommend on indigenous peoples' rights, no formal recommendations were made.[9] In sum, despite Cranbrook's best intentions, the Mission and the Report were a whitewash. The Report's conclusion that "sustainable management of the forests of Sarawak is being partly achieved," tended to delegitimize the burgeoning environmental campaign against Sarawak in Europe and North America. In not making recommendations regarding identified abuses of the legal process in Sarawak, the report failed the Penan and other indigenous peoples. And in making an AAC determination grounded in a negotiating rather than an environmental rationale, the report undermined the scientific foundations of the concept of "sustained yield."

Environmentalists were appalled with the outcome of the Sarawak Mission. The WWF Report of the November 1991 ITTO meeting at which the Sarawak Report was considered is scathing. It noted that "social issues were poorly analyzed and the final recommendations were weak"; that the report fails "to address the basic problems of recognition of customary land rights and reducing the overall cut rate"; that the decision to scale back production to 9.2 million m³ per annum over a five year period is "far too little and too slow" ; and that the U.S. efforts to have the Malaysian authorities report back to the ITTO on progress were not supported by other consuming countries who "did not push for it" (Elliott 1990).

From the perspective of environmentalists, the ITTO had systematically ignored their concerns and, worse, had used their presence at the ITTO to legitimize a set of unsustainable global policies for tropical forests that furthered industry and government objectives. As a result, they left the ITTO for other global fora where their interests would be less marginalized. And, unlike the 1980s, a large number of new fora were available, including the Convention on Biodiversity, the Commission on Sustainable Development, and the Intergovernmental Panel on Forests (later the Intergovernmental Forum on Forests).

The Mission to Sarawak constituted a watershed in the history of the ITTO. The Mission's report and specifically its recommendations established for the first time what the theoretical concept of "sustainable forest management" necessitated in practice. Neither industry, nor producing country governments, nor environmentalists liked what they saw. From the perspective of the Sarawak Government and timber industry, the reduction of the cut to nine million m^3 per annum was devastating, constituting a significant reduction to industry profits and government revenues. Thus, in accepting the Report's broad conclusion at the meeting of the ITTO, Sarawak and Malaysian officials were careful to point to factual errors in the report in relation to its calculation of the AAC. The official record of the meeting notes that "The Statement [of the State Government of Sarawak] took issue with some of the Report's findings and conclusions regarding the current rate of harvesting and the future of primary forests. It was the Government's view that the forecasts were based on excessive estimates of the consequences of harvesting" (ITTO 1990c: 19). At subsequent meetings of the ITTO, the Sarawak Government representative appeared regularly to account for why the State's AAC was increasing, not decreasing, in line with the Mission recommendations.[10]

Conclusion

A recent report by Poore and Chiew concluded that the ITTO has done more than any other international organization to promote SFM in tropically forested countries over the past 15 years (Poore and Chiew 2000). One can agree with that conclusion, without accepting its positive connotations. The ITTO promoted the concept of SFM as part of the solution to the world's deforestation difficulties. It is precisely the techno-managerial concept of SFM, however, that legitimizes continued, unsustainable levels of logging in Southeast Asia and elsewhere. SFM places governments and businesses as the center of the social institutions making national forest policy. It plays down the importance of civil society groups including ecologists, environmentalists and indigenous peoples' representatives. The discourse of SFM is resistant to alternative mechanisms to government regulation of forest practices—notably the potential efficacy of third-party ecocertification and ecolabeling.

The opportunity to witness the practical impact of the discourse of SFM occurred with the ITTO's Mission to Sarawak. The Mission's Terms of Reference made it difficult for non-techno-rational issues to be considered, such as the customary and legal land and forest rights of indigenous people. Yet even the techno-rationale approach of the Mission was compromised when Mission members could not agree on the appropriate size of the AAC and resorted to a process to negotiate the outcome.

The SFM discourse is still prominent in national and international negotiations on forest policy. It dominates the thinking of those governments and industry officials in Canada, Indonesia, Malaysia and the European Union that are pushing for a global forest convention. There has been a tendency for environmental and indigenous peoples' organizations to use the language of SFM while resisting its content. The experience of the ITTO suggests that this is a dangerous strategy. Resistance should occur not only on the ground in the forest, but in the rarefied air of scientific discourse, in the cities. Revealing the reductionist science of SFM paves the way for an alternative forest discourse to be put forward based on ecosystem principles. It is only through such a discursive shift that the balance between the use and conservation of forests in general, as well as of particular forests, will be achieved.

Acknowledgement

I would like to thank the Japan Center for Area Studies for organising such a stimulating, interdisciplinary conference and for inviting me to participate in its proceedings.

Notes

1 Thus, for example, the distribution of votes for the year 2000 gave 25 votes to each African country, with over 40 per cent of the total votes going to only three countries (Brazil received 158 votes, Indonesia, 136 and Malaysia, 130).

2 Japan continues to hold almost one third of the total votes with 323 votes out of 1000 in the year 2000 distribution of votes. The European Union has 293 votes and the other major consuming country, China, 156 votes.
3 The two other major programs where tropical deforestation was being dealt with were the FAO and the Tropical Forestry Action Plan (TFAP). Both were in disarray in the late 1980s. The FAO was unable to achieve much in the forestry sector, in part due to an ongoing dispute between its Executive Director and the United States over the organization's actions and budget. Meanwhile, the difficulties encountered by the Tropical Forestry Action Plan promoted by the United Nations Development Program, World Resources Institute and the World Bank created scepticism within government and civil society organizations over an exclusively aid-based solution.
4 Redclift states that the term's conceptual history can be traced to the Cocoyoc Declaration in 1974, its later usage by the Inter-national Institute for Environment and Development in the 1970s and its emergence as a key concept in the *World Conservation Strategy* published by the International Union for the Conservation of Nature in 1980. However, the term entered mainstream thinking only after the publication of the World Commission on Environment and Development's report, *Our Common Future*, in 1987 (for an account of the intellectual history of the concept, see Redclift 1987: 32–33).
5 Under section III of the draft, for example, Rietbergen included the clause: "Provisions for mandatory consultation with local peoples before road building and logging commences; continued exercise of customary rights; obligations for loggers to provide assistance, employment, compensation etc" (Rietbergen 1990).
6 Compare the language in the Rietbergen draft (see above) to the language in the ITTO guidelines. Principle 35 states, "The success of forest management for sustained timber production depends to a considerable degree on its compatibility with the interests of local populations." Principle 36 states, "Timber permits for areas inhabited by indigenous peoples should take into consideration the conditions recommended by the World Bank and the ILO in such areas *inter alia*" (ITTO 1992: 9).
7 In a ITTO Council discussion of the Report, one of its authors, David Cassells, is reported to have noted that "some practitioners felt that the ITTO Guidelines were outdated and that they should be updated to reflect the broader experience and the deeper conceptual understanding of sustainable forest management that had de-veloped over

the last decade." This very negative assessment is immediately qualified, however, because Cassells is reported to have noted also that "the majority of practitioners, however, felt that the existing ITTO Guidelines were imperfect but adequate and that the highest priority should be placed on implementation, monitoring and education" (ITTO 2000: 15). Whichever is the case, this is not a ringing endorsement of ITTO's Guidelines, either from the perspective of content or implementation.

8 The Brazilian delegation announced that his Government instructed him to place on the official record that "the Government of Brazil believed that it was a matter within the sovereign discretion of any Member State to request or welcome a mission to examine or evaluate its forest resources and formulate recommendations on forest matters ...Brazil considered the Resolution as 'sui generis,' a unique and exceptional initiative outside the competence of the ITTO which is a commodity organisation. Brazil did not recognise the competence of ITTO, through a resolution, to send a mission to a Member State or to establish investigatory procedures on national matters" (ITTO 1989c: 19–20).

9 The Mission Report notes that 'there is a sense in which it [the question of the traditional rights and claims of native communities to and in forested land] could be very much its [the Mission's] business. This is the extent, nature and location of these rights or claims and how, when and where they are exercised affect either or both the area and productivity of the forests' (ITTO 1990b: 8).

10 Thus, for example, as late as the 14th Session of the ITTO in 1993, the Chief Minister of Sarawak addressed ITTO delegates for the fourth time to defend Sarawak's implementation of the Mission's recommendations (ITTO 1993: 4).

Bibliography

Brosius, J. P. 1999. Green Dots, Pink Hearts: Displacing Politics from the Malaysian Rain Forest. *American Anthropologist* 101(1): 36–57.

Castilleja, G. 1990. Reducing the Annual Timber Harvest in Sarawak. How Much? DC: U.S. National Wildlife Federation, October 15, 1990. Mimeo.

Chimni, B. 1987. *International Commodity Agreements: A Legal Study*. London: Croom Helm.

Cording, U. 1985. UNCTAD and the International Tropical Timber

Agreement. In *The Proceedings of the IIED Seminar on the International Tropical Timber Agreement*. London: IIED.
CSSMM (Congressional Staff Study Mission to Malaysia). 1989. The Tropical Timber Industry in Sarawak, Malaysia. Report of a Congressional Staff Study Mission to Malaysia, March 25–April 2, 1989 to the Committee on Foreign Affairs, September. U.S. House of Representatives. Washington, DC: U.S. Government Printing Office.
Dudley, N., J-P Jeanrenaud and F. Sullivan. 1995. *Bad Harvest? The Timber Trade and the Degradation of the World's Forests*. London: Earthscan.
Elliott, C. 1990. Memorandum: ITTO Meetings 16–23 November, Yokohama, Japan, WWF Note on ITTO "Sarawak Mission Report." Gland, Switzerland: WWF–International.
——— 1993. Report on the 13th Session of the International Tropical Timber Council, 11th Session of the Permanent Committees and First Preparatory Meeting of the Renegotiation of the International Tropical Timber Agreement 1983 Yokohama, Japan 11–24 November 1992. Gland, Switzerland, WWF-International, February 20, 1993.
Gale, F. 1998. *The Tropical Timber Trade Regime*. London and New York: Macmillan and St. Martin's Presses.
——— 2000. Regulating Accumulation, Guarding the Web: A Role for (Global) Civil Society? In F. Gale and M. M'Gongile, eds., *Nature, Production, Power: Towards an Ecological Political Economy*, 195–214. Cheltenham, U.K.: Edward Elgar.
——— 2002. Caveat Certificatum: The Case of Forest Certification. In K. Conca, M. Maniates and T. Princen, eds., *Confronting Consumption*. Cambridge, MA: MIT Press, forthcoming.
Ghazali, B. and M. Simula. 1994. *Certification Schemes for All Timber and Timber Products*. Yokohama: ITTO.
——— 1996. *Timber Certification in Transition: Study on the Development in the Formulation and Implementation of Certification Schemes for All Internationally Traded Timber and Timber Products*. Yokohama: ITTO. 1998.
Government of Malaysia. *Forestry and Timber*. Kuala Lumpur: Government of Malaysia.
Hecht, S. and A. Cockburn. 1989. *The Fate of the Forest: Developers, Destroyers and Defenders of the Amazon*. London: Penguin Books.
Hpay, T. 1986. *The International Tropical Timber Agreement: Its Prospects for Tropical Timber Trade, Development and Forest Management*. Gland, Switzerland: IUCN.

IIED (International Institute for Environment and Development). 1985. *The Proceedings of the IIED Seminar on the International Tropical Timber Agreement.* London: IIED.

――― 1988. Natural Forest Management for Sustainable Timber Production. *ITTO Pre-Project Report* 11(88). London: IIED.

ITTA (International Tropical Timber Agreement). 1984. *International Tropical Timber Agreement, 1983.* TD/Timber/11/Rev1. New York: United Nations.

ITTO (International Tropical Timber Organization). 1989a. Labeling Systems for the Promotion of Sustainably Produced Tropical Timber. PCM, PCF, PCI (V)/1. Yokohama, Japan: ITTO.

――― 1989b. Incentives in Producer and Consumer Countries to Promote Sustainable Development of Tropical Forests. PCM, PCF, PCI (V)/1Rev. 2, Yokohama, Japan: ITTO.

――― 1989c. Draft Report of the International Tropical Timber Council on its Sixth Session. Yokohama, Japan: ITTO.

――― 1990a. Report of the Working Group on Guidelines for "Best Practice" and Sustainability in the Sustainable Management of Tropical Forests with Particular Application to the Natural Forests. Yokohama: ITTO.

――― 1990b. Report submitted to the International Tropical Timber Council by Mission Established Pursuant to Resolution 1 (VI) The Promotion of Sustainable Forest Management: A Case Study of Sarawak, Malaysia. Yokohama, Japan: ITTO.

――― 1990c. Draft Report of the International Tropical Timber Council at its Ninth Session. Yokohama, Japan: ITTO.

――― 1992. *ITTO Guidelines for the Sustainable Management of Natural Tropical Forests.* Yokohama, Japan: ITTO.

――― 1993. Draft Report of the International Tropical Timber Council at its Fourteenth Session. Yokohama, Japan.

――― 1999. Draft Report of the International Tropical Timber Council at its Twenty-Seventh Session. Yokohama, Japan: ITTO.

――― 2000. Draft Report of the International Tropical Timber Council at its Twenty-Eighth Session. Yokohama, Japan: ITTO.

Keto, A., K. Scott and M. Olsen. 1990. Sustainable Harversting of Tropical Rainforests: A Reassessment. Paper presented at a workshop held in conjunction with the Eighth Session of the International Tropical Timber Council, May16–23, 1990, Bali, Indonesian, Mimeo.

LEEC (London Environmental Economics Centre). 1993. *The Economic*

Linkages between the International Trade in Tropical Timber and Sustainable Management of Tropical Forests. London: London Environmental Economics Centre and the International Institute for Environment and Development.

Litfin, K. 1994. *Ozone Discourses: Science and Politics in Global Environmental Cooperation.* New York: Columbia University Press.

Maser, C. 1988. *The Redesigned Forest.* San Pedro, CA: R.E. Miles.

Nsenkyiere, E. and Simula, M. 2000. Comparative Study on the Auditing Systems of Sustainable Forest Management. Yokohama, Japan: ITTO.

OFI (Oxford Forestry Institute). 1991. Pre-Project Report on Incentives in Producer and Consumer Countries to Promote Sustainable Development of Tropical Forests. ITTO PPR 22/91. Yokohama, Japan: ITTO.

Poore, D. and Chiew, T. 2000. Review of Progress towards the Year 2000 Objective. ITTC (XXVIII/9), Yokohama, Japan: ITTO. http://www.itto.or.jp/ inside/report.html#review

Redclift, M. 1987. *Sustainable Development: Exploring the Contradictions.* London and New York: Routledge.

Rietbergen, S. 1990. A Synopsis of "Best Practice" Guidelines for Tropical Moist Forest Management for Sustainable Timber Production. London: IIED and ODA. Mimeo.

Sesser, S. 1991. A Reporter at Large: Logging the Rain Forest. *The New Yorker* 27: 42–67.

SKEPHI and R. Kiddell-Monroe. 1993. Indonesia: Land Rights and Development. In M. Colchester and L. Lohmann ed., *The Struggle for Land and the Fate of the Forests,* 228–263. Penang, Malaysia: World Rainforest Movement.

Utting, P. 1993. *Trees, People and Power: Social Dimensions of Deforestation and Forest Protection in Central America.* London: Earthscan.

UNCTAD (United Nations Conference on Trade and Development). 1977a. Consideration of International Measures on Tropical Timber: Report Prepared Jointly by the Secretariats of UNCTAD and FAO. TD/B/IPC/Timber/2. Geneva: UNCTAD.

——— 1977b. Report of the Preparatory Meeting of Tropical Timber Held at the Palais des Nations, Geneva, from 23–28 May 1977. TD/B/IPC/Timber/3. Geneva: UNCTAD.

——— 1977c. Report of the Second Preparatory Meeting of Tropical Timber held at the Palais des Nations, Geneva, from 24–28 October 1977. Geneva: UNCTAD.

——— 1978. Report of the Third Preparatory Meeting of Tropical Timber held at the Palais des Nations, Geneva, from 23–27 January 1978. Geneva: UNCTAD.

——— 1981. Draft "Articles of an International Agreement on Tropical Timber" Submitted by the Government of Japan, June 1982. TD/B/IPC/Timber/AC.2/Misc.4. Geneva: UNCTAD.

WRM / SAM (World Rainforest Movement and Sahabat Alam Malaysia). 1989. *The Battle for Sarawak's Forests.* Penang, Malaysia: WRM and SAM.

10
The Problem of Gaizai:
The View from Japanese Forestry Villages

John Knight

Over three-quarters of wood consumed in Japan is imported wood, known as *gaizai*, and much of that is tropical timber from Southeast Asia. North American temperate woods dominate the remaining kinds of *gaizai*, recent imports of which has caused much of the local discontent discussed in this paper. The wood trade has led to deforestation in much of Southeast Asia, but it has also had environmental effects in upland Japan. The trade links forests in Southeast Asia, America, and Japan itself, and transforms them. Taking a rather different approach from the other papers in this volume, this paper examines the impact of the Asia-Pacific wood trade in an area that is not commonly considered in tropical forest discussions, upland Japan—both the mountain villages and the forests (of timber plantations) that surround them. In the depopulated mountain villages of Japan, *gaizai* is viewed as a major cause of recent decline. *Gaizai* has displaced domestic timber from the market and accelerated the neglect and abandonment of timber plantations. This paper describes and examines the view of *gaizai* among the population of upland Japan, with specific reference to timber-growers. It draws on ethnographic fieldwork (during the late 1980s and 1990s) on the interior of the Kii Peninsula in western Japan and focuses, in particular, on an upland municipality in Wakayama Prefecture.

Japanese Mountain Forests

Japanese mountain villages are surrounded by forest. The 25 million ha of forest and woodlands make up around two-thirds of the national land area (Umebayashi and Oya 1993: 203). The verdant character of this landscape is captured in a variety of Japanese expressions describing it, such as *midori no rettō* ("green archipelago") and *shinrinkoku* ("forest country"). In Japanese, forest is known as *yama*—which means both "mountain" and "forest." The Japanese *yama* comprises two main kinds of indigenous forest: beech and Lucidophyllous forests. Beech forest (*bunarin*) is a cool temperate deciduous forest that includes Japanese beech (*Fagus crenata*) and is widely found in eastern Japan and at higher elevations (700–1,500 m). Lucidophyllous forest (*shōyōjurin*) is a warm temperate forest common in western Japan and at lower elevations (below 500 m) and also occurs in the wider East Asian area. This hardwood forest consists largely of evergreen oaks of the genera *Quercus*, *Castanopsis* and *Lithocarpus*, and is distinctive for its glossy broadleafed foliage (Kira 1995: 6). The Kii Peninsula is part of the Lucidophyllous forest zone of western Japan, and it is this vegetation that, until recently, surrounded most human settlements of the interior.

Around 14 million ha of the Japanese forest (56 per cent) is classified as "natural forest" (known as *tennenrin* or *shizenrin*) but this category includes primary and secondary forests. Of these, only an estimated five per cent of the forested areas are primary forests. Much of the "natural forest" is in fact pine forest that was created by human disturbances of the evergreen oak forest, like logging for timber or clearance for swidden farming (Nakagoshi 1995: 91). The red pine tree *Pinus densiflora* has long been exploited in Japan (and other parts of East Asia) for, among other things, timber, firewood, and pine resin. This secondary pine forest has been maintained under conditions of intensive human use; it includes deciduous oak trees that grow under the pine canopy and are cut for charcoal on a 20-year cycle (Kamada and Nakagoshi 1996: 20). The pine forest was maintained by regular human "disturbance activities" which had the effect of clearing the forest floor and allowing pine seedlings to establish. As a result of human withdrawal from the pine forest, this activity has been reduced and pine succession inhibited, thus allowing deciduous oaks beneath the pine canopy to eventually succeed the old pine stands (Kamada and Nakagoshi 1996: 24).

Japanese Forestry

Forestry has long been the dominant industry in upland Japan. Japanese forests have been manipulated to supply fuelwood and timber, with the preferred pattern of growth varying from one period to another. Since the 1950s, when modern forms of energy like electricity, oil, and gas replaced charcoal and fuelwood, forestry production has switched from hardwood

Photo 10-1. A family timber plantation in Wakayama prefecture, Japan Photo by Lye Tuck-Po

trees for charcoal to softwood timber species for housing, construction, and civil engineering. Timber plantations make up more than four-tenths of Japan's forest area and over a quarter (27 per cent) of the total land area. These plantations are dominated by two main tree species, Japanese cypress (*hinoki, Chamaecyparis obtusa*) and Japanese cryptomeria (*sugi, Cryptomeria japonica*), that are favored on account of their fast growth and adaptability to a range of different environments. Geometrical blocks of plantation forest, consisting of neat lines of standardized, same-aged trees, dominate the visible landscape of upland Japan. In the mountainous interior of the Kii Peninsula, timber plantations make up around two-thirds of the forest area.

The background to the present-day ubiquity of timber plantations is the large-scale logging that took place during the prewar and wartime periods, which resulted in a national landscape of extensive bare mountainsides. Overfelling had major environmental consequences in the form of landslides and extensive flooding in the downstream areas of Japan. One priority of the postwar years was to re-stabilize the national landscape by restoring tree cover to the mountains. A nationwide reforestation movement was launched after the war, with the Emperor as its figurehead, in which tree-planting events and ceremonies were held throughout the country. In 40 years, an estimated 10 million ha of new timber forest were planted and the national landscape of green mountains was restored. The postwar reforestation of the mountains has been officially proclaimed a great patriotic achievement on the part of the Japanese people and is a source of considerable pride within the Japanese forestry industry. In restoring trees to the mountains, reforestation replenished the nation's wood stocks and contributed to the recovery of national resource security. Japan's forest area has actually increased over the twentieth century as a whole, from around 22.5 million ha in 1900 to the present level of more than 25 million ha (Umebayashi and Oya 1993: 203). During this same period, the plantation forest area has increased tenfold to its current level of 10 million ha—from less than five per cent of the total forest area in 1900 to over 40 per cent in the present-day (*ibid.*).

Timber plantations on the Kii Peninsula generally date back to the 17th century, to the Yoshino tradition of regenerative forestry. Yoshino forestry is known for its highly intensive methods, including close planting of plantation saplings, frequent thinning of the growing trees, and long rotations to produce high-grade narrow-ringed timber. However, postwar plantation forestry differs from prewar plantation forestry in important ways. Before the Second World War, timber plantations were largely confined to north-facing mountainsides. The underlying idea that conifers such as cryptomeria grow better on the colder, less exposed north side, where the soil retains more moisture, is expressed by the traditional saying *yama wa kitamuki* ("the forest faces north"). The south side of the mountain consisted of a mixed forest, particularly of trees used for charcoal (like evergreen oaks and live oaks) that were believed to be made harder by the sun (Nomoto 1990: 25). Another feature of earlier plantation forestry was that a belt of mixed forest was left at the top of the mountainside near the ridge, while on high peaks a timberline of 800 m was generally observed. All this changed in the postwar period as a result of the reforestation program. Cypress and cryptomeria plantations were established on both sides of the mountains, south as well as north, while the new timber plantations often extend up to the ridge of mountains and in some cases beyond the 800 m line.

Japanese forestry is a labor-intensive industry. Forestry labor consists of both wage and family labor. On the large estates forestry is a permanent occupation, and there are even instances of second-generation forest laborers who work on estates that their fathers worked on before them. Although both men and women work as forest laborers, almost all full-time forest laborers (i.e., those who work over 150 days a year) are men, while women generally work less than half this number or days. Women are often employed in planting saplings in the spring and weeding young plantations in the summer.

In recent decades, private foresters have banded together to manage their forests, forming "Forestry Co-operatives" (*shinrin kumiai*) that employ forest laborers on which Co-operative members can draw for specific forestry tasks. It is estimated that these Co-operatives "now manage more than 70 per cent of all non-national forests in Japan" (JOFCA 1996: 13). They serve an important function because most individual foresters cannot afford to retain their own laborers and are not therefore in a position to undertake the harvesting of their plantations themselves. There is an increasing trend to subcontract even planting and silvicultural work to the Co-operative. In these decades, there has also been a sharp decline in the number of people working in the forestry industry. In 1960 there were 440,000 forest laborers in Japan, but by 1995 this number had fallen to just 80,000, a decline of over four-fifths (JOFCA 1996: 13)! The consequence of this labor shortage, on top of the trend towards absentee forest ownership, is a greatly diminished human presence in the timber plantations.

Forestry in Japan is in large part a family undertaking. In Japan there are two main forms of forest ownership: national and private. The *kokuyūrin* (nationally-owned forests) make up 31 per cent of the forest areas while the *shiyūrin* (privately-owned forests) make up 59 per cent. Of the 2.9 or so million private forest owners in Japan, some 2.5 million are individual forest owners (families) while the rest are companies, municipalities, and other public bodies (JOFCA 1996: 12). Individual forest owners "typically manage small-scale operations covering less than 5 hectares" (*ibid.*). The distribution of private forestland in Japan is highly unequal, as the postwar land reform, which redistributed farmland, did not affect forest landholdings. Full-time forest laborers are employed on the larger private forest landholding but in other cases it is the family members themselves who undertake forestry labor (apart from felling).

Representations of Timber-Growing

The Japanese stem family, the *ie*, is an intergenerational structure which ideally outlasts its individual members. The family thus consists both of extant family members and past generations of family members: the *ie* ancestors ritually memorialized in the domestic ancestral altar. The *ie* is a longitudinal entity that is perpetuated over time. It remains an influential ideal, regardless of the realities of actual family lives today. In upland areas, forest landholdings have been an important prop to the stem family. The family forest is an economic stake that helps to ensure family continuity through succession; as such, it is important to keep intact the forestland itself and pass it on down the generations. This over-riding duty is captured in the following proverb: *Binbō shitara mazu kazai o ure, tsugi ie o ure, tabatake o ure, sanrin o urubekarazu* ("If you are poor, sell-off first the family belongings, then the house, and then the farmland, but you should not sell the forest").

Part of the reason for the ideal of retaining forestland is that families become intimately bound up with the forests they own. The fruits of the labor of the earlier and present generations should be passed down to following generations in the form of good, sturdy trees. *Kodomo no tame ni birin o nokose* ("Leave behind beautiful forest for the children") is a proverb that gives expression to the family forester's "parental" obligation. Plantation forests require regular inputs of labor—something expressed in another proverb, *yama no hiryō wa waraji* ("sandals are the fertilizer of the forest").

It is because of this ancestral input of labor, together with the long time span involved in timber growth, that trees come to serve as an important medium for intergenerational family relationships. Foresters commonly point out that it takes three generations to grow good timber: the first generation does the planting, the second generation carries out the main silvicultural tasks that ensure fast, straight, and sturdy growth, while the third generation does the felling. But if it is the third, the harvesting generation, which benefits from the efforts of the earlier generations, it is incumbent on these beneficiaries to ensure the well-being of future generations by raising seedlings, planting saplings, and looking after maturing stocks that will only be felled much later.

Growing trees is commonly likened to raising children. As one local expression puts it, *yama wa kodomo o sodateru tsumori de* ("treat the forest as though you are bringing up a child"). Foresters explain this expression by pointing to the importance of the first ten years of a tree's life and the necessity of giving proper care to the young tree. The verb used for growing the tree is *sodateru*, "to raise," the same as that used for a child. It is the forester's responsibility to make sure that the tree grows the right way—that it is tight-ringed and knotless—as well as to protect it from hostile natural forces such as typhoons and snow, and from various unwanted kinds of plant growth and animal attention. Successful nurturance of trees, as with children, depends on regular "parental" contact and the taking of measures to remove potential threats to good growth.

An indication of the feelings foresters hold for their timber plantations is indicated by the comments of a 12th generation forest landowner on the Kii Peninsula, whom I interviewed in 1995.

> With the trees I've planted myself, or those trees planted by my father that I am tending to, I tend them because there is a feeling of love [*aijō*]. Along with that love, I feel that I must try and raise the trees [to maturity]. And so with each tree I've planted, after two or three years I have the sense that "ah, you've become this big." Each tree is different...You want to grow good trees that can then be sold—to raise and leave behind good trees for later descendants. I think there is this sort of cycle. Forestry is passed on from your grandfather, and then from your father, and on to yourself.

Two kinds of feelings for the trees seem evident in these comments. First, there is the (parental) "love" which the forester feels directly for the tree he himself has planted and raised. Secondly, there is the sense of obligation, to both ancestors and descendants, to ensure that good trees are grown, which can eventually be sold for good prices.

Among forest landowners, timber trees are often associated with the ancestor who planted and mostly tended to them. An example of this is the K family from one of the interior municipalities of the Kii Peninsula. At the end of each year, the 60-year-old family head sits before the family altar (*butsudan*) and reports on the plantations felled during the previous year. Most of the felling that occurs is of forests planted either by his grandfather or great grandfather (his father died young). On this end-of-year occasion, he expresses his *kansha* ("gratitude") to these ancestors for their efforts in making his present-day livelihood possible, and asks for their blessing to ensure that no injuries occur when tending the family forests over the coming year.

This man is, however, concerned about the future. His three children have all migrated to Tokyo. In an attempt to ensure that one of his two sons eventually returns to take care of the family forests, he built a wooden guesthouse in 1985, believing that, together with the forests, it would provide a stable economic base. The guesthouse was built entirely with wood from the family forests—the main pillars and beams of the new building were cypress trees planted by his great grandfather, while for the rest of it family cryptomeria trees were used. The father recalls that when reporting the felling of the guesthouse trees to the family ancestors, he gave particular thanks to his great grandfather for having planted the stands used, and to his grandfather and father for having raised them. But if forest landowners find family continuity to be a source of inner satisfaction, they are also deeply concerned whether that continuity can be secured in the future. In an era of large-scale depopulation, most forest-owning families on the Kii Peninsula are worried about family succession; many express apprehension over the prospects for the family forest in the absence of one of their sons returning from the city to become the family *atotsugi* ("successor"). One of the main effects of this trend has been a deterioration of the timber plantations.

It was estimated in 1995 that some 70 per cent of Japan's timber plantations were under 35 years old and therefore in need of silvicultural attention (JOFCA 1996: 3; cf. Itō 1994: 42). However, owing to the costs and scarcity of forestry labor, basic silvicultural tasks such as pruning and thinning have not been carried out in many of these plantations: around 40 per cent of timber plantations according to one estimate in the early 1990s (*Economist* 26/5/1992). Government figures on forest inventories give a further indication of the scale of the shortfall in thinning operations in recent decades. In 1995 the average volume in Japanese timber plantations was 182 cubic m per ha, whereas the forecast figure (for 1996) made in 1976 was 136 cubic m per ha, indicating a surplus of wood per plantation ha of 46 cubic m (Fenton 2001: 61). The effect of not thinning these plantations is to slow the rate of growth of the individual trees and, therefore, to extend the rotation time. To be sold for the sawnwood market, plantation trees must reach a diameter that permits the sawmiller to produce squared sawnwood columns of 10.5 and 12.5 cm (*ibid.*), but unthinned plantations will take that much longer to attain this target diameter.

Villagers find such neglected plantations a depressing sight to look at, and refer to them variously as *moyashiyama* ("bean sprout forest"), a forest resembling so many sprouts of a soya bean, and *senkōrin* ("incense forest"), a forest crowded with thin trees resembling so many incense-sticks stuck into the ground. Mountain villagers tend to blame the government for this all-too-visible deterioration in the timber plantations that surround them. They believe that the Japanese government, which encouraged foresters to plant their mountains over with timber conifers in the 1950s and 1960s, has a moral responsibility to support the sons and grandsons of the tree-planters today—but one which it has failed to honor.

Foresters' Views of *Gaizai*

In the course of post-war recovery, Japan experienced a shortage of usable wood products, and from the 1960s began importing large quantities of tropical wood from Southeast Asia. In subsequent decades the scale of Japanese wood imports has increased enormously: in 1965 imported timber accounted for only 13.3 per cent of timber consumed in Japan, but by 1994 this number had risen to over 75 per cent (JOFCA 1996: 8), while it has since increased to around 80 per cent. Japan imports around a third of internationally traded wood products. A high proportion of these imports take the form of raw logs that are then processed in Japan to conform to the specific requirements of the construction industry and furniture manufacturers. Most imported tropical timber is converted into plywood and about half of this plywood is used to make concrete molding cases (known as *konpane*) by the construction industry. Imported wood is also used for the pulp and paper industry—around 40 per cent of all wood consumed in Japan.

In recent decades, however, there has been a major shift. Japanese timber imports are now largely from temperate forests in North and South America, Australia and Russia. Of the 112,324 cubic meters (in roundwood equivalents) of timber and timber products imported by Japan in 1996, North American timber or *beizai* accounted for 38 per cent, whereas Southeast Asian and other sources of tropical timber accounted for less than 15 per cent (Blandon 1999: 87). Moreover, it is the import of coniferous timber from North America and elsewhere, which directly affects the Japanese construction industry and competes with domestic timber sources. Japan imports nearly a third of the coniferous logs that are traded in international markets, and purchases around two-thirds of the coniferous log exports from the United States (Blandon 1999: 90). In recent years competition between *beizai* imports and domestic timber has intensified and the marketing of imported construction timber has become much more aggressive. "Many Japanese companies are now importing *beizai* and marketing it as a product that is superior in strength and durability to the domestic species" (Blandon 1999: 100). It is the particular market threat presented by these coniferous timber imports, rather than by tropical timber, that largely informs the sentiments of Japanese foresters towards *gaizai* outlined below.

Japan is widely criticized for its high levels of wood imports. Japan is variously characterized as "the world's forest eater," "the world's greatest contributor to global deforestation," an "eco-outlaw," an "environmental predator," and charged with "crimes against the Earth." Environmentalist groups such as Friends of the Earth and Rain Forest Network have mounted campaigns, including the boycott of Japanese products, against the Japanese corporations involved. As the primary importer of tropical timber from Southeast Asia, Japan is condemned for causing widespread deforestation in the region. Japan is said to cast an "ecological shadow" over Southeast Asia in the form of assorted negative environmental impacts: "Together, Japan's ecological shadow and southeast Asian patron-client politics create a context that supports and accelerates destructive and illegal logging, contributes to ineffective reforestation and conservation polices, and undermines sustainable timber management" (Dauvergne 1997: 3). Invoking this same metaphor, it is of course the case that the Japanese "ecological shadow" is not confined to the tropical forests of Southeast Asia, but extends to the temperate forests of North America and elsewhere. Indeed, Japan is sometimes charged with cynically depleting overseas forests in order to conserve its own forests (Laarman 1988: 160–161; Myers 1986: 297–298).

In Japan too there is widespread disquiet at the high levels of wood imports. A number of environmental NGOs in Japan mounted a domestic anti-tropical timber campaign, one effect of which was to encourage local governments in Japan to limit the use of tropical timber in public construction projects (Wong 1998). There is also considerable opposition in the Japanese forestry industry to these high levels of wood imports. In upland areas of Japan the criticism of the international timber trade is that it has depressed domestic forestry by displacing from the market home-grown timber (supplying only half the amount of timber it did in 1965) and that domestic forests (i.e., plantations) have deteriorated due to neglect and abandonment (Knight 1997). Not only does imported timber displace domestic timber, it also depresses timber prices overall. There would appear to be strong support within the Japanese polity for decreasing the proportion of imported wood products in favor of domestic wood. In addition to reducing Japanese dependence on overseas wood supplies and reducing environmentalist criticism, such a policy would also provide a boost to the domestic forestry industry and revitalize this industry in remote mountainous areas, many of which have suffered large-scale depopulation. However, Japan's room for maneuver is limited by pressure from the United States and other trading partners which demand continued access to the Japanese market under the terms of the World Trade Organization.

What of the views of the foresters who grow the timber? How do they view *gaizai*? In Japan the distinction between *gaizai* ("foreign wood") and *naichizai* or *kokusanzai* ("domestic wood") is widely made. Many people use the *naichizai-gaizai* distinction as a simple dichotomy between two clearly contrasting categories. *Gaizai* is often stated to be weaker and less durable over time than *naichizai*, and as having an inferior appearance. However, as we shall see, there are also foresters and others connected to the forestry industry who qualify this distinction and point out that some kinds of *gaizai* are better than others, just as some kinds of *naichizai* are better than others. One common point of differentiation in the category is *beizai* ("American timber"), which many a forester and sawmiller admits can be of high quality and superior to other sources of softwood timber, whether domestic or foreign.

S is a 12th generation forest landowner who has 15 ha of forestland, around 12 ha of which are timber plantations (half of them cryptomeria and half cypress), and the remaining 3 ha *zatsuboku* or mixed tree species. Like most other forest landowners, he is angry with the government for allowing the present high level of wood imports. Local forestry is being "sacrificed" to international trade relations between Japan and America. The more industrial goods Japan exports, the more wood it is forced to import to reduce the trade gap! On the other hand, he then states that much of the criticism of *gaizai* voiced by local people should not be taken too seriously. In Japan, he pointed out, the further into the countryside you travel, the more you hear about the superiority of *naichizai* (domestic timber) over *gaizai* for building, but this talk is largely *uso* ("lies") by people who want to sell their own wood. He knows from experience that there is some very good foreign wood around, such as some of the North American timber, which can even be stronger and more reliable than domestic Japanese timber. Moreover, his experience in Southeast Asia tells him that Japan cannot compete on price. There are new timber plantations being established in places like Vietnam where the cost of labor is one-hundredth that of Japan. He is certain that the Japanese forestry industry will never be able to compete on price, only on quality. The challenge for Japanese foresters in the future is therefore to grow the best timber they can. But this is an undertaking that takes decades, whereas in modern society the concern is to make money now.

S also holds that *naichizai* is better suited to the Japanese climate. The principle is that of tree-to-wood compatibility: that wood for houses should come from trees from the forest. Foresters such as S advance this argument by invoking the teachings of the famous Japanese carpenter Nishioka

Tsunekazu, whose ideas are widely found among foresters, sawmillers and carpenters of the southern Kii Peninsula. One of the influential ideas articulated by Nishioka is that the specific pattern of growth of trees is directly connected to the durability of the buildings they subsequently make up. When the subject of wood imports arises, foresters often invoke a sort of nationalist bioregionalism, pointing out the unsuitability of wood that has been grown in a foreign climate. Foreign timber, it is often claimed, will not endure for very long in Japan's complicated, seasonal climate. Even though such timber may be cheaper than Japanese timber, in the long term this will prove to be a false economy.

> In Japan the climate is extremely variable. Because there is such a great variation in nature, the Japanese have believed since long ago that trees raised in America should be used in America, trees raised in Europe should be used in Europe, and trees grown in Japan should be used in Japan...For trees, unlike human lives, and are extremely tied to the climate of the place. So something that originates in that country is best kept in that country. (Nishioka and Kohara 1978: 76-77)

Nishioka invokes the same logic to argue for a sort of bioregionalism within Japan itself. The timber of one region should not be used for buildings in another region. Local wood, grown in the local climate, will make for the best building timber in that region. To use wood from outside the region is to use wood unused to the idiosyncrasies of the local climate, and therefore less likely to last.

These bioregionalist ideas are evident in the following comments of an official of the Shingū Timber Growers' Association made in an interview in 1995:

> The Japanese have long believed that, say, trees grown in America should be used in America, that trees grown in Europe should be used in Europe, and therefore that trees grown in Japan should be used in Japan. This is the best way forward I think...The reason for this is that, unlike the lives of people, trees greatly vary according to the temperature and climate. If they have been created in this or that country, it is best that they be used in that country...For example, it would be mistaken for tree species from Europe to be brought to Japan and planted here...Or if you took Japanese cryptomeria or cypress over to Europe or America, they would not grow properly there. It is the same thing as that...It has been only 30 years or so since foreign timber has been imported into Japan to build houses and so on...So as there is no real history yet to go on, you don't have proof that it is strong or it is weak...In growing my own forests I am grateful to the trees,

grateful to the mountains and grateful to nature. It is because of these sorts of feelings that I think that if I were to use foreign [wood] products, I would be offending my ancestors [*senzo san ni mōshiwakenai*]...That's all there is to it, and there is nothing complicated about it. It is not a case of hating foreign wood or anything like that...So, for Japan's foresters, when they see more than 70 per cent of wood being imported you would think that this is really unfortunate for them, but in fact this is not the case. In fact, they are grateful. What I mean by this is that after the war the demand for housing increased sharply and Japan did not have enough wood to supply the market, and so we asked various foreign countries and started importing their wood. During this time, we have been growing our own wood so that we now have a large stock of wood. If at that time, foreign countries had not helped us, all of Japan's [wood] resources would have been used up.

This man is opposed to the present-day scale of wood imports, believing both that it goes against nature and is at odds with the intergenerational contract with the ancestors. But in addition to this bioregionalist sentiment that leads him to condemn wood imports, there is also a pragmatic reasoning present in these remarks that leads him to offer qualified support—indeed "gratitude"—for wood imports on the grounds that they have protected Japan's forest "resources" during the period of high economic growth. Other foresters on the Kii Peninsula that I interviewed voiced a similar sentiment. The following comments are those of one of the largest forest landowners on the peninsula, a man who also owns much property in the peninsular tourist sector.

I believe that [wood] imports allow the Japanese forest to survive. Usually, for those people who work in the forestry sector, when imports increase their difficulties also increase, and so they desperately call for imports to be stopped. Although I cannot say in a loud voice to all those people who work in the wood-growing sector that "imports are also good," I do fear that if we did not know import wood the Japanese forest would gradually be cut down and our green environment would be lost. So although it may be bad for some of those in the wood industry, for consumers in the rest of Japan of course, they can buy [wood] cheaply. I think that if we say that our business is failing because of imports, what this means is that we are not trying hard enough. I think you have to ensure that you try sufficiently hard that you can succeed despite the imports. You have to make the effort to make sure that Japanese timber is not defeated and conquered by imports...I say to everyone I meet that even if cheap imports are coming in, what matters now is using your wits to find some way forward without being defeated by the situation.

Other foresters on the Kii Peninsula are resigned to *gaizai*. The T family has been doctors for many generations, and the current family head claims to be the 38th generation head of the family. He has lived and worked in Tokyo for 40 years, first as a hospital doctor and then as a professor of medicine. The Ts are one of the largest landowners on the southern Kii Peninsula, with a forest landholding of 500 ha. As the family head lives and works in Tokyo, the T forests are managed by a local *yamaban* or forest manager. During his return-visits to the Kii Peninsula, the family head will tour the estate with the *yamaban* and learn about the latest developments: which plantations have been felled, which have been replanted, the progress of young plantations, the readiness for felling of others, and so on. On one of these return-visits I spoke to him about his forests and raised the question of *gaizai* with him.

> Foreign wood is much cheaper to extract. I think it will continue to come into Japan in the future. In the Japanese mountains mechanization is difficult and the cost of employing people to work in plantations and extract the timber is really high. So when you think about the international competition we face, we have a real disadvantage. But, on the other hand, I am not one of those people who think that Japanese cryptomeria and cypress are always better than foreign wood.

This last remark contains implicit criticism of the views of many other foresters who assert the superiority of Japanese timber to foreign timber and claim that *gaizai* is weak and does not last.

Others who work in the wood industry also challenge such simple generalizations about domestic timber and foreign timber. The following remarks are from a member of the Shingū Timber Growers' Association:

> If you compare the domestic wood of Japan with foreign wood from outside, then in relation to Japanese architecture it is better to use Japanese wood. In any case, it is customary [in Japan] to use timber that you can obtain nearby...But it is still wrong to say that all foreign wood is bad. So you might have a situation where, compared with a 20- or 30-year-old domestic timber, it might be far better to use a foreign timber that would be far stronger and that would not differ much in price. What they do in the Yoshino area of Nara Prefecture, because they do not have access to long timbers locally, is to use American pine for everything above the pillar because of its strength. In the places that you can actually see, they use cryptomeria and cypress so that it looks nice, while in the hidden places that you can't see, that take all the strain, they use American pine, as this comes in the form of large timbers. But it is wrong for people to go around with

preconceptions that all foreign timber is weak and bad. You just have to differentiate. In the countryside people will insist on [native] cryptomeria and cypress, but in the cities they are not prized that highly.

In the Japanese regions there are frictions between foresters and sawmillers over the *gaizai* issue. In the interior municipalities of the Kii Peninsula one hears much criticism of sawmillers in the coastal cities who opt to use foreign timber. By contrast, sawmillers in the interior, who actually live in forestry villages, tend to use domestic timber sources. U, born in 1949, is one of these sawmillers. On finishing high school he left for Tokyo where he worked as a lighting technician in a Shinjuku theatre, but in 1977 he returned with his family to live in his natal village on the Kii Peninsula. His family has significant forest landholdings, and U knows many forest landowners as well as forest laborers. He attends the various timber markets on the peninsula. Like most people, U is critical of Japan's high level of wood imports. His sawmilling business only handles wood from local forests. He would not consider handling "foreign timber" or *gaizai*: "I have never used even one foreign log." He is adamant that, while it may be very cheap, foreign timber is unsuited to Japan's climate and therefore inappropriate for use in Japanese buildings.

Unfortunately, some sawmillers in Japan, including those on the coast, do handle *gaizai* and make much money from doing so. From a forest landowning family himself, U believes that it is very sad that there are local sawmillers who no longer support local forestry. Many sawmillers have prospered by using cheaper imported wood, while wood-growers, seeing timber prices flatten because of the imports, have suffered relative decline. The center of gravity of the regional wood industry has therefore appreciably shifted in favor of the sawmillers, while tree-growers have seen their power over the wood industry decline. Accordingly, forest landowners' feelings towards sawmillers are decidedly mixed: the sawmillers are seen as benefiting from the imports that undermine the landowners' own market position. The only way forward is to form an alliance with sawmillers to try to present a united front in favor of a ceiling on imports.

Conclusion

As Japan enters the 21st century, it finds that many of its erstwhile industries are moving overseas. Part of this trend is the overseas translocation of wood growing, as with other primary, and secondary, industries. One consequence with respect to timber forestry is that upland space in Japan comes to be assigned new land-uses, such as tourism—witness the boom in forest recreation in recent years (Knight 2000). But this economic process is also bound up with what might be called issues of moral economy. Although timber is a commodity, grown for the market, it is also symbolically infused by the families that grow it and the national purpose it is deemed to meet.

The sentiments of the Japanese foresters outlined above should be understood, to a large extent, in terms of the unfavorable market position in which they find themselves. They see their own timber being displaced from the domestic market by large-scale timber imports, especially from North America. What I have referred to above as the "bioregionalist" rhetoric should be understood, to some extent, in this connection, and recalls similar Japanese producer rhetorics that emerged in connection with the liberalization of other sectors, such as the agricultural markets for rice and beef in the 1980s and 1990s. Yet there is no doubt that the above representations of timber growing also express a profound dissatisfaction with the way the international market for timber undermines the local social relations of timber production. Upland spaces in Japan have historically provided a sense of continuity. As the landscape changes and people move out, this sense of continuity may come to serve a purely symbolic function, for the practical activities that link upland and lowland, supplying wood for lowland construction, have steadily been eroded.

In order to understand the full implications of forest conversion, it may be insufficient just to focus on those places where logging activities are most current. This account has shown what happens after the logging stops. It is not just about the cessation of a particular set of economic activities, but also concerns transformations in how people see themselves, their roles in society, and their identity both within a household and in the country more generally. At the same time, pressuring governments like Japan's, as is done by environmental groups, to consider the impact of logging industries on forest communities in Southeast Asia and other places, may only be a small part of the solution. This discussion shows that domestic concerns too must be a critical part of the policy shift through which sustainable forest management is successfully undertaken. Political ecology conventionally examines how resource use is embedded in the disparities and conflicts engendered by changes in the political economy. The lesson from Japan's upland timber-growing regions is that the source of conflicts may have deeper roots and wider implications than we normally credit them.

Bibliography

Anon. 1992. Symposium: Who Will Care for the Forest? *Economist* 26/5/1992.

Blandon, P.R. 1999. *Japan and World Timber Markets*. New York: CABI Publishing.

Dauvergne, P. 1997. *Shadows in the Forest: Japan and the Politics of Timber in Southeast Asia*. Cambridge: The MIT Press.

Fenton, R.T. 2001. Forecasts of Forest Products Demand in Japan. *International Forestry Review* 3(1): 58–63.

Itō, M. 1994. Mokuzai Shigen to Jinkōrin" (Timber Resources and Artificial Forestry). In F. Konta, S. Watanabe and Y. Takei, eds., *Shizenrin no fukugen (The Restoration of the Natural Forest)*. Tokyo: Bunichi Sōgō Shuppansha, 41–44. (With enclosed English-language translation).

JOFCA (Japan Overseas Forestry Consultants Association). 1996. *Forestry in Japan*. Tokyo: JOFCA (Supervised by the Forestry Agency).

Kamada, M. and N. Nakagoshi. 1996. Landscape Structure and the Disturbance Regime in Three Rural Regions Hiroshima Prefecture, Japan. *Landscape Ecology* 11(1): 15–25.

Kira, T. 1995. Forest Ecosystems of East and South-East Asia in a Global Perspective. In E.O. Box, ed., *Vegetation Science in Forestry: Global Perspective Based on Forest Ecosystems of East and South-East Asia*, 1–21. Dordrecht: Kluwer Academic Publishers.

Knight, J. 1997. A Tale of Two Forests: Reforestation as Discourse in Japan and beyond. *Journal of the Royal Anthropological Institute (MAN)* n.s. 3(4): 711–730.

———— 2000. From Timber to Tourism: Recommoditizing the Japanese Forest. *Development and Change* 31(1): 341–359.

Laarman, J. 1988. Export of Tropical Hardwood in the Twentieth Century. In J.F. Richards and R.P. Tucker, eds., *World Deforestation in the Twentieth Century*. Durham, N.C.: Duke University Press.

Myers, N. 1986. Economics and Ecology in the International Arena: The Phenomenon of "linked linkages." *Ambio* 15: 296–300.

Nakagoshi, N. 1995. Pine Forests in East Asia. In E.O. Box, ed., *Vegetation Science in Forestry: Global Perspective Based on Forest Ecosystems of East and South-East Asia*, 85–104. Dordrecht: Kluwer Academic Publishers.

Nishioka, T. and J. Kohara. 1978. *Hōryūji o Sasaeta Ki (The Trees Offered Up to the Horyuji Temple)*. Tokyo: NHK Books.

Nomoto, K. 1990. *Kumano Sankai Minzokukō (A Treatise on the Mountain and Coastal Folk Customs of Kumano)*. Kyoto: Jinbun Shoin.

Umebayashi, M. and K. Oya. 1993. Participatory Forest Development and Management: The Japanese Experience. *Regional Development Dialogue* 14(1): 202–218.

Wong, A. 1998. The Anti-tropical Timber Campaign in Japan. In Arne Kalland and Gerard Persoon, eds., *Environmental Movements in Asia*, 131–150. London: Curzon.

Index

access 4, 5, 107–130
a*dat see* customary laws and practices
Agathis borneesis 155
agriculture and farmers 1, 9–10, 13–14, 29–65, 79, 84, 86–87, 95, 122, 155, 159; cash crops 30, 86, 123, 177–197, 224; colonial planters 13; government subsidies 190, 192, 197; smallholders and peasants 6, 19, 58, 97, 121, 124, 138, 140–142, 177–197 *see also* swidden agriculture
agroforestry *see* forestry
Albizzia 52; *A. molucanna* 44
Alnus nepalensis 52
Amazon 6–7
animals 76, 78, 83, 88–91, 100–101, 108; animal products traded 31, 81–83, 87–92, 95, 98–100; habitat reduction and overhunting 90–91
appropriation and expropriation 4–5, 14–16, 18–19, 108
aristocrats *see* elites
Asian financial crisis (1997) 77–78, 121, 147, 221–222
autonomy 19, 121, 161, 169 *see also* regional politics and development; micro-politics

Bagobo of Mindanao 56
bamboo 20, 36, 41, 52, 107; products 78, 120
Banjarese of Kalimantan 136, 138
benefits and benefit-capture 2, 4, 16, 29, 95–98, 107–130, 159, 224 *see also* access; appropriation; coercion; forest products and resources; resource regimes; trade
benzoin 72, 83, 89, 93, 98, 100
biodiversity 2, 7, 17, 21, 31, 33, 43, 47, 58, 61, 76, 79, 87, 90–91, 134–135, 152, 201, 205, 211–214, 224
bioregionalism 276–277, 280
birds nests caves 12, 112, 114–115, 156, 158
Borneo 7, 12, 20, 47, 110–111, 113, 136, 157 *see also* Sabah; Brunei; Kalimantan; Sarawak
botanical gardens 35, 46–47, 49, 51, 53
Brazil 240–241, 243, 245, 258, 260
British *see* colonialism and colonial regimes
Brunei 111, 136, 180, 183

Bugis 110–114, 116, 138, 156, 158
Burma (Myanmar) 30–31, 33–36, 38, 41–44, 51–53, 58, 61–62, 76, 80–82

Cambodia 76, 79–80, 82, 94
camphor (*Dryobalanops* spp.) 32, 50, 155
carbon sequestration 209–211, 214, 220, 224
cardamon 72, 83, 87–89, 93, 98
cash economy and -earning opportunities 78, 91, 109, 114, 123–124, 155, 182–197, 268–269
CBRM *see* natural resource management; resource regimes and conflicts
certification and eco-labeling 222–223, 226, 244
chengal (*Balanocarpus maximus*) 59
China 12, 17, 19–20, 74, 76, 78–82, 84, 91, 93, 95, 110, 111, 158; vassal states 82
Chinese 48, 85, 94, 110, 113–114, 158
Chinese medicines 82, 87
Christianity 155
CITES 90–92
claims 117, 122, 182 *see also* conflicts
climate and deforestation 31, 35, 37–38, 40, 43–45, 48, 50–51, 57, 61–62, 205, 209–210, 212, 224 *see also* carbon sequestration
cloves 135

coconut 138–139, 142, 144, 146, 152; products 139
coercion: actions and measures 5, 18, 107–109, 112, 117, 120, 122–125, 127–128
coffee 37, 44, 85–86, 89, 91, 155, 238
cogons see grasslands
collusion 14, 108, 110, 125, 127, 129, 164
colonialism and colonial regimes 11–13, 17, 21, 29–32, 36, 46–47, 49, 57, 61–62, 72, 74–75, 80, 84–86, 89, 91, 96–97, 110–111, 113–115, 124–125, 128, 155, 157–158 *see also* India; forestry; Burma; Indonesia; Laos; Malaysia; Philippines
conflicts 4, 15–16, 18, 72–101, 107–130, 152–173, 219–220, 236–260, 265–280; micro-politics 153, 165, 168 *see also* cooperation; property rights; resource regimes; territorial issues
connections and influence 4, 9, 14, 119, 128, 169
conservation 2, 8, 13, 107, 124, 134–136, 149; legislation 89, 91 *see also* NGOs; protected areas
cooperation 5, 18, 110
corruption 118, 146, 165 *see also* coercion; collusion; cronies; rents
cost-benefit analysis *see* values and valuation
costs 16, 108, 134, 200–226

INDEX

criminalized acts 8, 40, 60, 76, 117, 123 *see also* rents
cronies 5, 8, 14, 98, 119, 121, 146, 149
cultivation *see* agriculture; swidden agriculture
customary laws and practices 9, 113–114, 116, 121, 127, 159, 164, 167–168, 182; *tanah ulen* 115, 124–125 *see also* claims; elites; land; rights and responsibilities; property rights

damar 83, 156, 172, 180
Dayak 12, 15, 112, 125, 152–153, 155–156, 158–159, 161–162, 164–165, 167–168 *see also* Iban
decentralization 15–16, 18–19, 120, 122, 125, 129, 152–173
deforestation *see* environmental conditions and changes; forest management
demand and supply 13, 18–20, 54, 60, 62, 72, 78, 84–87, 89, 91, 93, 95, 97–98, 100, 108, 110, 114, 119–120, 157, 160, 190–191, 202–204, 220–223, 265–280
development 8–9, 72–101, 107, 122, 133–149, 160, 200–226; trade-offs with conservation 152, 200–226
devolution *see* forest management; regional politics and development
Dipterocarp forests and species 2, 14, 29, 33, 47, 54, 60, 62, 155, 207–208 *see also* forest types
disasters and disruptions 94, 194 *see also* fires
disciplinary approaches 3–4
Dutch *see* colonialism and colonial regimes

ecology 2–3, 21, 200–226
economic conditions and changes 4, 12, 16, 22, 87, 96, 121, 128, 147, 177–197, 200–226 *see also* cash economy; demand and supply; development; regional politics and development
economic theories 3, 5, 109, 200–226; neo-classical 16, 214–216
elephants 89, 91, 100
elites 8, 11–12, 41, 74, 79, 85, 96–97, 99, 108, 110–117, 125, 128, 135, 149, 152, 156, 158–160, 164, 169
El-Nino 135
environmental conditions and changes 1–7, 9–10, 13–15, 17–18, 29–30, 40, 52, 72, 87, 89, 98–99, 107–108, 121, 125, 134, 138, 142, 146–147, 149, 177–197, 200–226 *see also* laws and policies
environmental groups *see* international linkages and negotiations; NGOs
environmental ideas and theories 6, 29, 31–32, 35, 38

exclusionary practices 38, 41–43, 49, 115 *see also* access; appropriation and expropriation; coercion; government administration
exploitation *see* access; coercion; logging; resource regimes
exports *see* demand and supply; prices; trade; trading networks and routes
extraction *see* forest products and resources; trade

Federated Malay States *see* Malaysia
fires 1, 11
Foochow Chinese 15
forest clearance, conversion, and degradation *see* environmental conditions and changes
forest diversity and ecology *see* biodiversity
forest management 10, 18–19, 22, 30, 41–43, 45, 48, 53, 58, 62, 139, 236–260; devolved 20, 124–125, 134, 153–173 *see also* natural resource management; resource regimes; sustainability
forest products and resources 2, 4–5, 8–9, 12, 16, 19, 30, 41, 44, 47, 58, 72–73, 75, 79, 82, 84, 87, 91–92, 95–97, 99, 107–108, 111–116, 120–122, 124, 128, 155–156, 169, 177–197; as basic needs 42, 44, 49, 52, 60, 76–78; availability and viability 89, 91; collection 12, 16, 78, 95–96, 98–99, 108–109, 115–116, 119–120, 123, 177, 180–181, 183, 186, 189, 193–194, 196; collectors 8, 59, 108, 110, 114, 125, 157; concessions 117; development of 133; exploitation 112, 116; extraction 110; industries 128; procurement 109, 113; producers 123, 125, 127–128; protectors 110 *see also* benefits; demand and supply; natural resource management; resource regimes; trade
Forest Reserves 30–31, 34–35, 38, 41, 42, 44, 49, 50, 52, 53, 56, 58, 60–61, 164; "standard tree" 30
forest types: dry evergreen 76; hill and mountain forests 21, 58, 76, 265–280; Japanese beech forests 265–280; Japanese Lucidophyllus forest 265–280; lowland rain forest 51; logged-over 11; mangroves 35, 50, 52–53, 59, 64, 138; natural forest 44; non-teak forests 45, 58; peat swamp forests 2, 16–17, 19, 21, 55, 58, 65, 133–149, 152; primary forests 108, 161, 164, 182–184; subtropical broadleaf forest 76; swamplands 177–197; temperate woods 202, 265–280 *see also* uplands
forest-dependent communities 5–6, 9, 11, 13, 18, 52, 89, 96, 134–135, 167–168

forestry 3, 29–62; administrations and agencies 13, 16, 29–31, 33, 35–36, 39–40, 45–47, 49–51, 53–57, 61–62, 75, 96; agroforestry 52; annual allowable cut (AAC) 237, 255–258; as family enterprise and symbol of continuity 265–280; community forestry 121, 153, 165, 167–168, 170; foresters 13, 18, 21, 30–31, 36, 50–51, 53, 62, 244–246, 255–256, 265–280; legislation 12, 14, 33–34, 36, 38, 40–41, 43–44, 51, 58, 116–117, 127, 160, 168; scientific forestry 30–33, 36, 40, 44; silviculture 41, 62; decline 269–280 *see also* laws and policies; multilateral institutions and linkages
fuel 43, 49, 77, 201, 221, 266; reserves 52–53
futures 10–11, 22, 134, 168, 206, 209, 215–216, 218, 220–223, 270–272, 275, 278

gaharu (eaglewood; *Aquilaria* spp.) 85, 87, 95, 120, 125, 155, 156, 158
gambier 47–48
gambodge 83, 85
girdling 31, 43, 58
global markets *see* demand and supply
government administrations and representatives 4, 9, 11, 15, 18, 34, 37, 47, 49–50, 56, 72, 82, 89, 91, 96, 110, 113, 116–118, 121, 125, 127–129, 133, 139, 161, 164, 167–168, 180, 177–197 *see also* land
grasslands 37, 54, 56, 182–183, 186–188, 191, 194, 196
gutta percha 32, 47, 50, 53, 59, 112, 158; *Palaquium* spp. 32, 47; *Payena* spp. 32, 47

hunting *see* animals
hydro-electric dams 10–11
hydrology 133–134

Iban 16, 21, 177–197 *see also* Dayak
Imperata see grasslands
incomes *see* cash economy
India 31, 33, 36, 38, 41, 43–44, 46–47, 49–51, 58, 61, 101
indigenous groups 7, 107, 121, 127, 134, 160, 164, 167, 168, 246–247, 253–256, 258–259 *see also* inland groups; local communities; uplands
Indochina 72–101
Indonesia 2, 6–7, 11–16, 19–20, 30–31, 33, 36–39, 41, 44–47, 51–55, 57–58, 61–62, 65, 85, 107–129, 133–149, 152–173, 202, 204, 207–208, 213, 221, 223, 239–241, 243, 245, 251–252, 258; *korupsi, kolusi, nepotisme* 121; *orde baru* 117, 125, 127, 129; PIR project 139–142; PLG rice project 136, 144–148; post-Suharto era 121, 167; *Reformasi* 121, 125, 129 *see also* decentralization; Borneo; forestry; Java; Riau; Suharto

infrastructure 33, 60, 85–87, 95, 98, 112, 177, 180–182, 186, 189–195, 197, 200–201, 206, 209
inland groups 12, 112–113, 127 *see also* forest products and resources; indigenous groups; uplands
interest groups 109, 128
interethnic relations 152, 164 *see also* conflicts
international linkages and negotiations 6, 17, 77, 148, 200–226, 236–260 *see also* NGOs; multilateral institutions
investments 72, 78, 85–86, 91, 94, 98–100 sust, 161
ironwood 115
Irrawaddy delta 35, 43, 51–53, 82
ITTO (International Tropical Timber Organization) 16, 18, 22, 236–260
ivory 81, 85

jade 82
Japan and Japanese markets 19–20, 120, 200, 202–204, 211, 221–222, 238–241, 243, 259, 265–280; perceptions of wood imports 265–280
Java 30–31, 33, 36–39, 41, 44, 46, 51–55, 57–58, 61–62, 111, 113
Javanese 14, 44, 114, 138
jelutong (*Dyera* spp.) 32, 58, 65, 183

Kachins of Burma 51–52
Kalimantan (Indonesian Borneo) 6–7, 11–12, 15–16, 19–20, 58, 111, 113–116, 118–125, 136, 145–146, 148, 153, 156–159, 169–170 *see also* Malinau
Karen of Burma 34–35, 37–38
Kayan 115
Kenyah 112–113, 115–116, 124–125, 153, 156, 159, 162–163; Kenyah Lepo'ke 155, 159
keruing (*D. cornutus*) 155
kings and kingdoms 11–12, 41, 74, 79–80, 84, 110–117, 125, 128, 156, 158 *see also* elites
Korea 200, 203, 221

land 7, 37, 42–43, 46, 110, 113, 117, 121–122, 134, 139, 161, 180–181 *see also* customary laws and practices; Forest Reserves; government administrations; laws and policies; property; resource regimes; rights and responsibilities; tenure; territorial issues
Laos 16, 72–101
laws and policies 6, 8, 14, 18, 37, 94, 99, 109–110, 113–114, 117, 120–121, 127–128, 161, 164, 177–180, 184, 190, 192–193, 196–197, 222, 238 *see also* forestry; forest management
leaders and leadership 6, 112, 115, 157–159, 161, 164–165, 168 *see also* elites; kings and kingship
local communities 6, 12, 16, 116, 124–125, 135, 139, 153, 162, 165, 168, 170; self-

INDEX 289

characterization of 164; as perceived 163–164, 167–168 *see also* indigenous groups; inland groups; uplandslogging 1–2, 7–11, 15, 17–20, 31, 35, 43, 50, 53, 56–59, 62, 86, 107, 116–118, 128, 134, 184–186, 189–191, 193, 195, 201–204, 206, 209, 218–219, 221–222, 224, 265–280; companies and concessions 8, 20, 46, 54–55, 99, 119, 129, 135, 149, 155, 160–162, 165, 186, 202–204, 211; logging boom 2, 10, 14, 184 *see also* benefits; revenues and royalties

Malacca 48, 50, 59, 95
Malay groups and states 12, 56, 60, 110, 114, 138, 158
Malaysia (Malaya) 2, 6–7, 12–17, 19–21, 30, 33, 36, 47–51, 54–56, 58–62, 72, 79, 95, 97, 110–111, 113, 120, 136, 157, 177–197, 200–204, 207–208, 213, 221–223, 239–241, 243–244, 248, 252–258, 260 *see also* Borneo; Malacca; Sabah; Sarawak
Malinau 15, 152–153, 155–157, 159–160, 162, 167
markets *see* colonialism and colonial regimes; demand and supply; forestry; logging
Mekong 75, 80, 82, 85, 94
Melanau of Sarawak 15
meranti (*Shorea* spp.) 155
Merap of Kalimantan 115, 124, 153, 156, 157–160, 162–164, 167

merbau (*Afzelia palembanica*) 59
merchants 6, 82, 84–85, 94–95, 97–98, 112, 179, 183, 185–187, 190–191, 197
micro-politics *see* access; conflicts; resource regimes
migrants and migrations *see* population movements
military 8, 11, 98, 125
mills 58, 86, 186, 189, 196–197, 272, 275–276, 279
minerals and mining 60, 155, 158, 160, 162, 165; gold and silver 84, 110; tin 59, 86, 238
molave (*Vitex parviflora*) 33, 39
monopolies 37, 85, 96–97, 135, 236
monopsony 125–127
moral economy 4, 15, 280
multilateral institutions and linkages 6, 90–92, 149, 222–223, 259; International Monetary Fund 221; World Bank 6, 75, 78, 97, 134, 245; World Trade Organization 78 *see also* international linkages; ITTO
Myanmar *see* Burma

nationalization of forests 152, 158, 160
natural resource management 3, 10, 59, 169, 177–197 *see also* access; coercion; forest products and resources; resource regimes
NGOs (non-governmental organizations) 6, 9, 18, 117, 124–125, 134, 148–149, 241, 243–246, 250, 252, 256–259

see also multilateral institutions and linkages
non-timber forest products (NTFPs) *see* forest products and resources

oil (petroleum), effects of 135, 181, 183, 189–190, 197, 237, 266–267
oil palm 1, 20, 89, 121–124, 138, 164; plantations and companies 20, 122–124, 133, 164; smallholders 124 *see also* agriculture; population movements
Orang Asli 59–61
ownership *see* access; property; resource regimes

paddy *see* agriculture; rice; swidden agriculture
panglong woodcutters' camps 55, 58, 65
Papua New Guinea 136, 200, 202–204, 208, 223
paradigms 21, 29–30
patronage *see* collusion; connections and influence; elites; leaders
patron-client relationships 15, 157, 160, 165
pearls 82, 84
peat lands *see* forest types
Penan *see* Punan
Penlay kanazu 36
pepper 49, 85
Philippines 6–7, 14, 30, 33, 38–41, 45–47, 53–57, 60–62, 84, 136, 200, 202, 204, 208, 221–222

plantations and their development 1, 7–9, 10–11, 13–14, 31, 36, 44, 51, 59, 61, 85, 89, 122, 139–142, 144, 265–280; crops 121, 127 nuclear estate and smallholders project 140, 141, 142, 144 *see also* agriculture; oil palm; population movements; rubber
plants and plant products 31, 77, 82, 85, 87, 108
political and economic control 34, 110–116, 124, 148, 157
political ecology as an approach 2–7, 9, 11, 15–18, 21–22, 29–30, 32–33, 74, 107–108, 153, 200–226; as applied to Asian forests 2–3
political economy as theory and approach 3–4, 6, 15, 201–207
political-economic inequities 2, 4–5, 220–226 *see also* power; poverty
population movements 138–139, 152, 156, 159, 167; immigrants 85, 138; organized immigrants 138; resettlement 76, 158–159; settlers and settlements 139, 167, 177–197; spontaneous migrants 138, 142, 144; squatters 60; transmigrants 6, 136, 139, 141, 146 outmigrations and depopulation 265–280 *see also* Indonesia
Portuguese 84, 114
postmodern approaches 6
poverty and wealth disparities 9, 78, 134, 201–202, 224

power 5, 11, 15–16, 18, 22, 109, 128, 135, 152; abuses of 5, 165; bargaining power 108; in decision-making 122, 164; in politics 5, 112, 128, 152, 159, 161, 162, 164, 169; inequitable distribution 4, 109; power relations 108; shifts and balances of 107, 123, 152; structures 5, 31
pre-colonial period 21, 110, 155, 125
prices 59, 86–89, 91, 94, 96, 98, 113, 117–120, 122–124, 126, 139, 146, 164, 166, 183, 185, 188–190, 192–194, 196, 202–203, 205–206, 211, 221, 223, 225, 237–238, 250, 271, 274–275, 278–279
property rights and conflicts 42, 96, 110, 112, 114, 116–117, 122, 127–128, 133, 236 *see also* conflicts; forest management; timber management
protected status and areas 57, 89, 134, 214 , 246
provincial governments *see* government administrations; regional politics and development
Punan / Penan 7, 15, 113, 115, 155, 156–157, 160, 162, 164, 168, 180–181, 253–254, 256; Foundation for Punan Customary Law 164, 168; Punan Malinau 155, 157, 159, 164; Punan Tubu 155, 157, 159, 161, 164, 168
Putuk of Malinau 153

rail systems *see* infrastructure
rattans 19–20, 47, 72, 115, 118, 120, 155–157, 180, 183, 192; rattan gardens 20, 118, 121, 122, 124, 157; producers and suppliers 118–119; rattan products 119, 189, 191–192, 194, 197; trade and demand 119, 123
raw materials 78, 81, 119, 123
reforestation 57, 145, 267–268
regional politics and development 16, 79, 117, 123, 141, 161, 165–166, 169; district governments and officials 127; provincial governments and officials 15, 110, 116, 127
rents 5, 94, 107–130 *see also* benefits; revenues and royalties
resins 47
resistance 44, 169, 194
resource regimes and conflicts 2, 4–5, 14–15, 17–18, 41, 44, 46, 52, 60, 98, 108, 110, 112–116, 121; community-based 46, 54, 114, 117, 121 *see also* conflicts; property rights; territorial issues
resources *see* animals; forest products and resources; logging; minerals and mining; plants; resource regimes and conflicts; timber
revenues and royalties 79, 82, 84–85, 91, 94–97, 111–112, 114, 123, 135, 162, 180
rhinoceros 89, 91; products 81
Riau 16, 55, 58, 65, 135, 135–142, 144, 149

Riau PIR-Trans project *see* Riau
rice 78–79, 84–86, 135–136,
    145–147, 155, 164, 186–188
rights and responsibilities 4, 42,
    85, 121, 127, 131, 155, 161,
    169, 182, 246–247, 253,
    255–256, 258–259, 265–280
Rio Summit (Earth Summit 1992)
    9, 201
risk minimization 112, 194
road systems *see* infrastructure
rubber 2, 7, 30, 33, 47, 49–51,
    59–62, 85–86, 89, 108, 122,
    178, 180, 184, 188–197, 238;
    smallholders and tappers 7,
    58, 72
rulers *see* elites

Sabah (British North Borneo) 33,
    55–56, 61, 120, 204
sago 155
Sambu of Riau 142, 145
sandalwood 110, 114–115, 117–
    118, 127; ownership and
    management 117, 125, 127
Sarawak 7, 14–16, 19, 21, 62,
    110–111, 113, 120, 136, 157,
    177–197, 202, 243–244,
    252–258, 260
scarcity and loss 87, 89, 109,
    146, 112, 180–181 *see also*
    species loss
science 3, 6, 30–31, 236–260 *see
    also* forestry; multilateral
    institutions
Shans of Burma 51
shipping 12, 96–97
SIJORI triangle 17, 138
Singapore 6, 17, 33, 48, 50, 54,
    138

slavery 113, 115
social relations 15, 152–173, 219
Solomon Islands 200, 202–204,
    207–208, 226
Soviet Union 80, 87
species loss *see* biodiversity
spices 82, 84, 87 *see also* trade
Srivijaya 111
stratification 99, 115, 165, 168
    see also control; power
sugar and salt 85, 183
Suharto, President of Indonesia 8,
    14, 19, 21, 121–122, 135–
    136, 144–149 *see also*
    Indonesia
sultans and sultanates *see* elites;
    kings and kingdoms
sustainability 100, 108, 134;
    sustainable development 2,
    200–226; sustainable forest
    management (SFM) 17–18,
    21, 237–260 *see also* forest
    management; natural resource
    management; resource
    regimes
swidden agriculture 1, 6, 9–10,
    13–14, 37–38, 40–41, 44, 52,
    77, 79, 117, 122, 155, 159,
    177–197; cultivators and
    farmers 13, 21, 32, 38, 40, 51,
    54, 56, 61, 124, 156; fallows
    56, 87, 182–183, 188, 190,
    194; plots 54; "robber
    farmers" 33, 44

Taiwan 200
Taosug of Kalimantan 112–113,
    116, 156
taxes and tributes *see* revenues
    and royalties

tea 89, 238
teak (*Tectona grandis*) 29, 31, 33–39, 52, 55, 58, 61, 62, 86; geographic locations of 21, 31, 34, 36, 38, 41, 44–45, 54, 61; regeneration of 37, 61; teak *taungyas* 34–35, 41, 52
technology 84, 94–96, 110, 184, 204, 223
tenure 44, 159
territorial issues 12, 19, 41, 44, 46, 52, 90, 111–113, 115–116, 125, 127, 165, 167–168 *see also* customary laws and practices; land; resource regimes and conflicts
Thailand (Siam) 75, 78–80, 82, 84–85, 91, 94–95, 97
tidal irrigation and water management (*pasang surut*) 138, 142 *see also* forest types
Tidung 156, 158
tigers 83, 88–89, 100
timber industry *see* forest management; forestry; logging
timbers 1, 7, 11, 14, 19, 29–65, 72, 91, 118, 121–122, 127–128, 149, 158, 162, 184, 191, 236–260 *see also* logging
trade 11–12, 72–101, 107–130, 180, 255; geographic areas of 84, 112, 127; items of 47, 73–75, 81–83, 91, 93, 99–100, 110, 112, 121, 129, 135; policies 119; private trade 85; spice trade 84; stratification of 99; taxes and duties levied on 85, 94, 111; traders 84–85, 96–97, 99, 113–114, 116, 119, 123, 125, 127; world agreements 237–238, 247, 249–250 *see also* benefits; collusion; demand and supply; monopolies
trading networks and routes 12–13, 87, 93–97, 111–114, 127
traditional organizations 110–113, 160, 165, 168 *see also* customary laws and practices

uplands 2, 12, 18–20, 30, 33–35, 37–38, 51–52, 57–58, 62, 72–100, 167, 265–280

values and valuation 4, 16, 29–30, 160, 240, 246–247, 200, 203–207, 214–215, 200–226, 265–280
Vietnam 46, 75–76, 78, 80, 84–85, 89, 91, 94, 98
Vietnamese 97
village leaders *see* elites; leadership
villages *see* access; cash economy; elites; forest management; natural resource management; resource regimes; trade; trading networks and routes
VOC 39

Wakayama Prefecture (Kii Peninsula) 265–280
wars and their effects 47, 75, 94, 110, 156, 177, 179–181, 183–185, 188, 190, 193; raiding 113; World War II 80, 85–86, 179–181, 183–185, 188, 190, 193, 202